C. Klett
Particle Size Distribution
in Fluidized Bed Systems

3-89927-002-9

ad libri

Time-Dependent Behavior of the Particle Size Distribution in Fluidized Bed Systems with Recirculation of Solids

Vom Promotionsausschuss der
Technischen Universität Hamburg-Harburg
zur Erlangung des akademischen Grades
Doktor-Ingenieur
genehmigte Dissertation

von
Cornelis Klett
aus Bad Oldesloe
2005

1. Gutachter: Prof. Dr.-Ing. J. Werther

2. Gutachter: Prof. Dr.-Ing. A. Kather

Tag der mündlichen Prüfung: 7. November 2005

Cornelis Klett

Time-Dependent Behavior of the Particle Size Distribution in Fluidized Bed Systems with Recirculation of Solids

adlibri Verlag

adlibri Hochschulschriften – Verfahrenstechnik

Die Deutsche Bibliothek – CIP-Einheitsaufnahme

Die bibliographischen Daten dieser Veröffentlichung sind in
Der Deutschen Bibliothek – Deutsche Nationalbibliographie –
verzeichnet und im Internet abrufbar (http://dnb.ddb.de).

adlibri Verlag GmbH & Co. KG
Postfach 13 01 91, D-20101 Hamburg
www.adlibri.de

Printed in Germany
ISBN 3-89927-002-9

Es gibt einen Fluch, der lautet:
Mögest Du in interessanten Zeiten leben.
(T. PRATCHETT, 1997)

Die Zeit als wissenschaftlicher Mitarbeiter und Doktorand am Arbeitbereich für Partikeltechnologie der Technischen Universität Hamburg-Harburg war für mich in der Tat sehr interessant in vielerlei Hinsicht. Oft genug habe ich mir gewünscht, sie könnte ruhig etwas weniger interessant sein. Aber ich habe viel gelernt, und dafür möchte ich an dieser Stelle denen danken, die dafür im Wesentlichen mitverantwortlich waren.

An erster Stelle danke ich meinem Doktorvater Professor Dr.-Ing. Joachim Werther für die Geduld, die er mit mir hatte und die Freiräume, die er mir einräumte. In erster Linie war er es, der – zusammen mit Ernst-Ulrich Hartge – dafür sorgte, dass aus einer Idee, die zunächst nur ein Hobby war, ein interessantes und anspruchsvolles Promotionsthema wurde und blieb.

Danken möchte ich auch Jens Reppenhagen. Er war es, der mich letztendlich ermutigt hat, mit der Promotion zu beginnen. Ich habe es nicht bereut.

Dass die Zeiten interessant waren und blieben, lag sicherlich auch an meinen Kollegen am Arbeitsbereich. Von diesen hat insbesondere Carsten Vogt, mit dem ich das Büro teilte, immer dafür gesorgt, dass es nie langweilig wurde. Zu Beginn meiner Arbeit haben Stefan Bruhns und Torben Edens mir in vielerlei Hinsicht beigestanden – später wurden diese hervorragend durch Claus Reimers und Reiner Wischnewski vertreten.

Bernd Schult und Heiko Rohde haben im besonderen Maße dafür gesorgt, dass die Versuche und Experimente so gut funktionierten. Auch von ihnen habe ich dabei vieles gelernt.

Ich bin sicher, dass die Zeiten nicht nur für mich interessant waren, sondern auch für die Diplomanden Jessica Schreiber, Patrick Engemann und Kristin Rosenkranz sowie die Studienarbeiter Claus Reimers, Michael Balfe, Petra Vilt und Tibor Hänsel. Das Gleiche gilt sicherlich auch für die vielen studentischen Hilfskräfte. Ohne die Genannten wären die umfangreichen Versuche und die dabei gewonnenen Erkenntnisse und Ergebnisse nicht denkbar gewesen.

Mein besonderer Dank gilt meiner Familie. Sie hat mir diese Möglichkeit geboten – und mir in meinem beruflichen und privaten Leben, sowohl in interessanten als auch schwierigen Zeiten, beigestanden.

Ob meine interessanten Zeiten für die übrigen Beteiligten und Betroffenen Segen oder Fluch waren, müssen diese selbst entscheiden. Für mich selbst kann ich sagen, dass sie – nun im Nachhinein betrachtet – ein Segen waren.

Dafür allen noch einmal von Herzen: Danke.

Table of Contents

Symbols

a	decay constant (cf. eqn. (3-58)), -
A	heat transfer surface area of probe, m²
A_{el}	Area in the cyclone defined by eqn. (3-34)
Ar	Archimedes number, -
A_t	cross-sectional area of fluidized bed, m²
A_w	clarification area in the cyclone, m²
b	exponent defined in eqn. (3-20) or in the "Gwyn-Equation" eqn. (2-1), -
C_b	particle size independent rate constant of bubble-induced attrition, s^2/m^4
C_c	particle size independent rate constant of cyclone attrition, s^2/m^3
C_j	particle size independent rate constant of jet-induced attrition, s^2/m^3
$c_{p,water}$	heat capacity of water, kJ/kg/K
c_v	particle volume concentration, -
$c_v{}^*$	theoretical particle volume concentration at infinite height, -
$c_{v,app.}$	Apparent particle volume concentration, -
$c_{v,be}$	particle volume concentration at the interface between bottom bed and upper dilute region, -
$c_{v,e}$	particle volume concentration at the riser exit to the cyclone, -
$\bar{c}_v$	average particle volume concentration in the freeboard of the riser, -
c_{vd}	particle volume concentration of the suspension phase in the bottom bed, -
C_w	drag coefficient of particles, -
D	parameter in separation efficiency curve, eqn. (3-37)
d_{50}	average particle size, m
$d_e{}^*$	cut size of separation at cyclone entrance, m
d_{or}	diameter of an orifice in a multihole gas distributor, m
d_p	surface mean diameter, m

d_p*	cut size of separation in the inner vortex of the cyclone, m
D_t	diameter of fluidized bed, m
d_v	diameter of bubbles in the fluidized bed, m
d_{v0}	initial diameter of bubbles in the fluidized bed at gas distributor, m
h	height above gas distributor, m
H_b	Height of bottom bed, m
h_e	height of cyclone inlet, m
H_f	Height of freeboard, m
h_{vf}	height of vortex finder in the cyclone, m
Δh	height segment of the riser, m
k	exponent in eqn. (3-28)
$k_{\infty,i}$	to mass fraction related entrainment flux of particles in size class i, kg/m²/s
K_a	attrition rate constant in the "Gwyn-Equation" eqn. (2-1), s^{-b}
k_b	heat transfer coefficient at probe surface, kJ/s/m²/K
m_{att}	mass of abrasion-produced fines per one time step, kg
$\dot{m}_{att.}$	mass of abrasion-produced fines per unit time, kg/s
m_b	bed mass, kg
$m_{bed,0}$	initial bed mass, kg
$m_{c,in}$	mass entering the cyclone within one time step, kg
$\dot{m}_{c,in}$	solids mass flux into the cyclone, kg/s
$m_{c,o}$	mass leaving the cyclone within one time step through the overflow, kg
$m_{c,u}$	mass leaving the cyclone within one time step through the underflow, kg
m_e	mass entrained from fluidized bed within one time step, kg
m_e	mass, which leaves the riser to the cyclone within one time step, kg
m_f	mass of solids in the freeboard, kg
m_i	material mass in the size interval i, kg

$\dot{m}_{i,i-1}$	mass transfer flux due to particle shrinking from the size interval i to (i-1), kg/s
m_{loss}	mass lost from a given system, kg
m_{tot}	total mass of solids in a given system, kg
$\dot{m}_{water}$	water mass flow rate, kg/s
n^*_c	passes through cyclone, where $\dot{m}_{att,c} = 1.1 \cdot \dot{m}_{att,c,\infty}$, -
n_{or}	number of orifices in a multihole gas distributor, -
ΔP	Pressure drop, Pa
ΔP_M	measured pressure difference of one height segment of the riser, Pa
$Q_3(x)$	cumulative particle size distribution in mass, -
$\Delta Q_{2,i}$	fraction of the size interval i on the entire particle surface, -
$\Delta Q_{3,i}$	fraction of the size interval i on the entire material mass, -
$\dot{Q}$	heat flux, kJ/s
r	Attrition rate, kg/kg/s
r_2	radius of cyclone at half of the conical part, m
r_a	inner diameter of cylindrical part of cyclone, m
r_e	average radius related for inlet flow into the cyclone, m
Re	Reynolds number of particles at fluidizing velocity, -
Re	Reynolds number, -
r_{vf}	radius of vortex finder in the cyclone, m
T	temperature of riser, K
$\overline{T}$	average temperature of water in probe, K
t	time, s
t^*	time, where $\dot{m}_{att} = 1.1 \cdot \dot{m}_{att,\infty}$, s
T_{in}	water temperature at probe inlet, K
T_{out}	water temperature at probe outlet, K
Δt	time interval, s

u	superficial gas velocity, m/s
u_b	rising velocity bubbles, m/s
$u_{c,in}$	gas velocity at the cyclone inlet, m/s
u_{mf}	minimum fluidization velocity, m/s
u_{or}	gas velocity in the orifice of a multihole gas distributor, m/s
$\dot{V}$	volumetric flow rate, m³/s
$\dot{V}_b$	visible bubble flow, m/s
w_s	terminal velocity, m/s
$w_{s,50}$	settling velocity of a particle with at cyclone with settling probability of 50%
x	particle size, m
x_i^*	particle size which will travel from one size class i to the next size class i-1 by shrinkage, m
$\overline{x}_i$	geometric mean particle size of the size interval i, m
$\overline{z}$	average centrifugal acceleration, m/s²
$\overline{z}_2$	average centrifugal acceleration of the flow for the first rotation of the flow in the cyclone, m/s²
z_{vf}	centrifugal acceleration at radius of vortex finder in the cyclone, m/s²

Greek symbols

ε_b	bubble phase fraction in the fluidized bed, -
η_F	fractional separation efficiency in inner vortex of the cyclone based on particle size, -
η_V	fractional separation efficiency in inner vortex of the cyclone based on terminal velocity, -
ϑ	stress history parameter, -
$\Delta\vartheta$	change of stress history parameter within one time step or pass through a cyclone, -
λ	splitting factor for bubble growth, -

λ_s	friction coefficient in the cyclone, -
μ_G	critical solids loading for the carrying capacity of the gas , -
μ_i	solids loading of size class i in inner vortex, -
μ_{in}	solids loading at the cyclone inlet, -
$v_{r,vf}$	radial velocity at radius of vortex finder in the cyclone inlet, m/s
$v_{\varphi 2}$	tangential velocity at characteristic radius r_2 in the cyclone inlet, m/s
$v_{\varphi a}$	tangential velocity at the cyclone outer radius, m/s
$v_{\varphi e}$	tangential velocity at average stream line in the cyclone inlet, m/s
ρ_f	density of the fluid, kg/m³
$\rho_{f,L}$	density of the fluid at ambient conditions, kg/m³
ρ_s	density of particles, kg/m³
ρ_{sus}	density of solids-gas suspension, kg/m³
$\Delta\rho$	density difference between solids and gas, kg/m³

Indices

0	value at initial state
∞	related to steady state condition
b	bed
c	cyclone
$exp.$	obtained in experiment
i	index to number a certain size interval
j	jet or index to number a certain stress history class
$loss$	lost material
or	orifice
p	for a single particle
tot	related to the entire system

Abbreviations

FCC	Fluidized Catalytic Cracking
PAPSD	primary ash particle size distribution
PSD	particle size distribution
SEM	scanning electron microscope

1. Introduction

In a lot of gas-solid processes fluidized bed technology is applied with great success. The advantages of fluidized beds which are namely good gas-to-solids mass and heat transfer, homogeneity of temperature and concentrations make them very convincing for application. They are applied for a lot of different purposes as for chemical reaction, granulation or coating of fine particles and for combustion in power plants or waste incinerators, respectively. With fluidized beds a high degree of product quality and / or reliability on low emissions can easily be achieved with the appropriate design of the process. Besides a lot of design criteria as geometry, flow rates, temperature, pressure, etc. the particle size distribution of the handled solids has a big impact on the performance and goes as well into the design.

The design of a process starts therefore often with the particle size distribution of the raw material, e.g. catalyst, fuel, etc. Mostly the particle size distribution of the raw material is given and decided by pretreatment, availability on the market and other factors. The particle size distribution of the delivered solids, which are fed to the process, does not have to be the particle size distribution, which will develop in the process. In fact the particle size distribution will even differ locally in the process significantly. The particle size distribution at a certain position or apparatus in the process depends on the previous process steps, performance of the apparatus itself and on the properties of the particles, which can be reactivity and friability. For the design of a process the particle size distribution has to be described for each process step in conjunction with all the other steps in the flow sheet and the individual process units have to be designed accordingly to meet the desired process specifications, e.g. mass flow rates, particle size distributions of product, conversion efficiency, energy efficiency, etc.

However the particles' size distribution is decisive for a lot of operating characteristics. In first place it decides the fluidizing behavior of the bulk solids (A, B, C or D according to GELDART (1973)). Fine particles tend to form a homogeneous fluidized bed with very small bubbles whereas coarser particles to be fluidized with a much more vigorous movement, big bubbles and spouting behavior at the same fluidizing velocity. Bubbles are a very important issue in fluidization. Bubbles are causing a good mixture of the solids in the bed (BELGARDT et al., 1987). Bubbles are distributing particles in the vertical direction by dragging them in their wake from the bottom to the top, where they are released again. To conserve mass balances this forms an inner flow field in the bed, which also generates a lateral movement of solids and influences their mixture in this direction. Consequently this affects the bed-to-wall heat transfer (MOLERUS 1992). On the opposite side big bubbles are allowing gas to bypass the bed and the gas might not contribute to the reaction, which decreases the conversion efficiency, e.g. in fluid catalytic cracking processes (WERTHER & HARTGE, 2004).

Besides the internal effects of the particle size distribution in a vessel the particle size distribution is decisive for the mass flows of solids to following process units. In fluidized beds the size of particles decides whether a particle is entrained with the leaving gas or remains in the vessel and might have to be discharged from the bed to keep a desired hold-up of solids at a steady level. In cyclones the entering particle size distribution and mass flow has to be considered for the design of the cyclone to obtain the correct separation efficiency. However this is valid for every other thinkable apparatus in the process whose design has to be according the entering stream.

But the particle size distribution is not only affected by separation and mixing. Furthermore it changes by degradation of the individual particles themselves. The particles are exposed to a vigorous movement in the system, which causes mechanical stress and leads to breakage and abrasion. This results in a change of the particle size distribution with time, when the solids are kept in the system as it is often realized in catalytic processes. In processes with a continuous feed of solids like combustion or calcination the change of the particle size distribution of the solids is influencing the mass fluxes between vessels in the system, the mass fluxes leaving the system and the product quality, respectively.

A lot of effects are influencing the particle size distribution of the material in a process and most of them have already been investigated and modeled. The foregoing investigations were dealing with single process steps or apparatuses. When a whole system has been investigated it was often regarded as a single unit ("black box"). Combining these models is not easy to get a general description of an arbitrary process. The purpose of the present work is to investigate the suitability of the existing models for the general description and to modify them accordingly to make them suit if necessary and to develop an approach for modeling where no model exists.

After reviewing the previous investigations in the state of the art (c.f. chapter 2) the strategy of the modeling of a complete fluidized bed system with recirculation of solids is described in the theoretical section (chapter 3). Here the dynamic modeling of an arbitrary combination of fluidized beds with cyclones is explained and the interfaces between the so-called modules are defined to provide the possibility to change the combination of apparatuses to fit an arbitrary process.

A clear focus is set on the attrition of particles, where it was found in previous works that the attrition behavior of the particles depends on their fate, which they have experienced before. A model is developed, which calculates the production rate of fines and shrinkage of particles depending on their previously experienced stress. In chapter 3 an approach is developed to describe the change of the attrition behavior in dependence of the so-called "stress history". It is described how the stress history is changing in the individual process units, which are the fluidized bed and the cyclone. The experimental investigation of the influence of the stress history on the attrition and the extraction of model parameters from the experiments is described in chapter 4 and the results are presented in the following chapter 5.

The second focus of the theoretical work is put on how a system can be described, which contains solids from different materials. The implemented models were therefore investigated regarding their potential to consider the different properties of solids from different materials and are adapted to meet the suitability. The adaptations, which are applied, are described in the theory (c.f. chapter 3).

Finally two complete systems of fluidized beds with recirculation of solid, were experimentally investigated and the results compared with calculations of the developed model to give a validation. The strategy here is to generate the needed model parameters in independent experiments and use them as input parameters for the simulation of the complete system. The measurements of the existing systems are then compared with the calculated results without any fitting of parameters to show the agreement with the previously made assumptions (c.f. chapter 6).

Finally predictive calculations are performed for some typical situations during operation of a circulating fluidized bed combustor, which are the addition of sorbents for emission control and the change in load, respectively. In these applications the model calculations are used to predict the change in the particle size distribution and the consequences of the changes for the process.

2. State of the art

2.1. The influence of the particle size distribution on the performance of fluidized bed systems

Besides operating conditions as gas velocity, pressure, temperature and the geometry of a fluidized bed system the particle size distribution of the bed material is an essential characteristic parameter. A lot of effects inside the system are strongly dependent on the particle size distribution. Some experiences of industrial applications which are directly related to the particle size distribution of solids will be presented in the following section. Afterwards the industrial observations and conclusions are supported by investigations made under laboratory conditions.

2.1.1. The Influence of the particle size distribution on industrial-scale operation

Most of published industrial experiences are made at fluidized bed combustors. Experiences regarding the operation of chemical reactors are very rare although a lot of scientific work is dealing with effects occurring in these systems. Therefore two examples are given here of experiences in combustors.

The operators of the Zeran power plant in Warsaw / Poland had recently the rare and unique opportunity to observe the influence of cyclone performance on the particle size distribution of the inventory of the circulating fluidized bed boiler and as a consequence on several operating parameters. The observations are presented by LALAK et al. (2003). When the power plant was extended by a second boiler it was decided to build an absolute identical boiler B besides the existing boiler A with only one difference in installing an improved high efficiency cyclone. All other parameters as geometry, operating conditions, fuel and bed material were kept the same. They observed first of all an increase of fines kept in the system which leads to a decrease of surface mean diameter from 180μm in boiler A to 80μm in boiler B. As a result the limestone utilization is significantly improved as well the bed to wall heat transfer coefficients. This leads to a significantly reduction in NO_x and CO emissions and power consumption for air supply.

Experiences at the circulating fluidized bed plant of Stadtwerke Duisburg AG (HEIDENHOF & ALTHOFF, 2003) show the consequences of change in fuel which changes the particle size distribution of bed material significantly. They stated that a change from a fuel rich in ash with a majority of coarse particles to fuel [from South Africa] which generated mostly fine ash particles the amount of bottom ash decreases to an extend that most ash left the system as fly ash. The lack of bottom ash has then to be compensated by addition of quartz sand. This leads in turn to an increase of erosion at heat exchanger surfaces due to higher hardness of sand compared to ash. The particle size distribution of the bed material is therefore decisive for the ratio of fly ash to bottom ash.

From applications in the chemical industry it is reported by DE VRIES et al. (1972), that an increase in the conversion of gaseous hydrogen chloride in the Shell Chlorine process from 91 % to 95.7 % was achieved by increasing the fines content in the bed material from 7 % to 20 %. The same effect was observed by PELL AND JORDAN (1988) with respect to the propylene-conversion during the synthesis of acrylonitrile. They reported on an increase of the conversion from 94.6 % to 99.2 % as the fines content was changed from 23 % to 44 %. Unfortunately experiences with the application of fluidized beds in chemical industry are fairly poor published. However these two examples demonstrate the importance of the particle size distribution in these processes as well.

2.1.2. The influences of the particle size distribution in laboratory-scale experiments

Besides the reported industrial experiences there are scientific researches on the influence of particle properties including their size on effects in application available in literature. CHENG et al. (2004) investigated the sulfur capture capability of sorbents at fluidized bed condition. They found the caption capability to increase with an increase of mass specific surface area. This corresponds to the experiences at Zeran (LALAK et al., 2003). The decrease of surface mean diameter will consequently result in an increase of mass specific area and the observation of better desulphurization can be explained.

With respect to influences of heat transfer some measurements were recently conducted and results were published. BREITHOLZ et al. (2001) correlated the measured heat transfer coefficients to the suspension density at the heat transfer surface. Since the distribution of solids and local solids volume concentrations in a fluidized bed riser are depending on the particle size distribution an indirect influence is indicated. Additionally to the suspension density SUNDARESAN & KOLAR (2002) correlated their results with the average particle size of the solids.

The conversion of gaseous compounds is also influenced by the particle size distribution of the bed material. According to HILLIGARDT & WERTHER (1987) the size of bubbles is decreasing with decrease in particle size. Consequently the mass transfer from bubble phase to suspension phase is increased (SIT & GRACE, 1981) as well the number of bubbles. As a result the conversion is increased as WERTHER & HARTGE (2004) could demonstrate in model calculations where the influence of cyclone performance on a FCC-reactor was investigated.

2.2. The factors which are influencing the particle size distribution in fluidized bed systems

The factors that are influencing the particle size distribution in a fluidized bed system with recirculation of solids can be divided into two groups. The first group of factors is degradation effects of the individual particles and the second one are separation effects, which are changing the particle size distribution by separating smaller from bigger ones. Degradation of particles is a big field and involves any kind of size reduction of particles as fragmentation, shrinkage and attrition. In the scope of this work only attrition in the meaning of abrasion is considered.

In figure 2-1 the distinction between the two degradation mechanisms attrition (here understood as being the same than abrasion) and fragmentation is sketched. Fragmentation leads to a broad particle size distribution as result where abrasion does change the particle size distribution slightly and produces a significant amount of fine particles without changing the size of the mother particles significantly. Therefore it takes a much longer time to reduce the size of particles to an elutriable size by abrasion than by fragmentation. Abrasion as attrition is the degradation mode, which will be in the focus of this work.

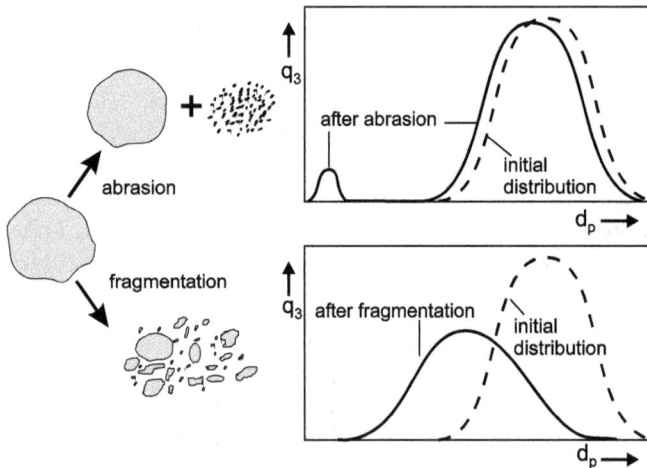

Figure 2-1: The definition of abrasion and fragmentation.

2.2.1. The Attrition of the particles in fluidized bed systems

Attrition has a very strong influence on the particle size distribution of solids in fluidized bed systems due to the vigorous movement of solids. Therefore it is investigated in parallel for many applications, e.g. FCC-catalyst, fuel and ash in combustion and at substitute particles to investigate general effects.

The attrition of FCC-catalysts causes operating costs due to loss of catalyst by attrited fines. Therefore a lot of work was focused on the development of less friable catalyst particles and techniques to classify them. Examples of these standardized tests are submerged jet tests (FORSYTH & HERTWIG, 1949 and GWYN, 1969) or the Grace-Davidson jet-cup tests (WEEKS & DUNBILL, 1990 and DESSALCES et al., 1994).

These experimental investigations were not applicable to predict the real material loss of catalyst under operating conditions, but to classify FCC and make their friability comparable. XI (1993) and REPPENHAGEN (2000) extended the experimental methods to measure the attrition rates of FCC under nearly operating conditions and the sources occurring in fluidized bed systems in isolation. In these investigations the main sources of attrition in fluidized bed systems were identified. XI (1993) investigated the attrition in the vicinity of gas jets at the gas distributor and the bubble-induced attrition within a fluidized bed. In the vicinity of gas jets which are penetrating the fluidized bed at the gas distributor, solids are accelerated by the incoming gas at high velocities and then colliding with the slower particles in the dense bed itself. In the bed rising bubbles are causing a vigorous movement of the particles with reasonable relative velocities, which causes friction and abrasion. The findings of XI (1993) were extended by REPPENHAGEN (2000) by the investigation of attrition, which happens in cyclones. When the particles are entering the cyclone they are colliding with the cyclone walls due to their inertia or are colliding with other particles in the swirling flowfield.

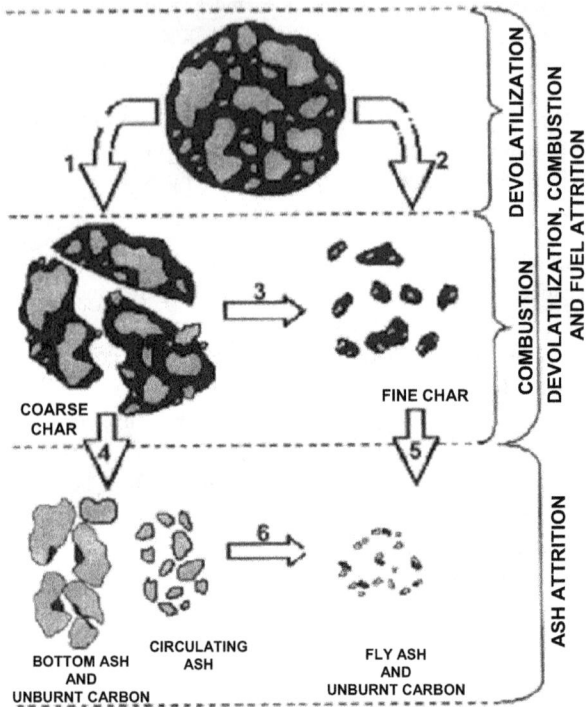

Figure 2-2: The mechanisms of ash formation in a fluidized bed combustor (from CAMMAROTA et al., 2002)

In contradiction to FCC-catalyst attrition the attrition of particles in a circulating fluidized bed combustor is much more complex. The bed material of a combustor does not only consist of one material. It consists usually of inert bed material, which could be a mixture of sand and ash. Additionally there is fuel inside and often also sorbents as limestone is added. In these systems several effects, which cause the size change of particles, are taking place at the same time.

CAMMAROTA et al. (2002) recognize 6 mechanisms (c.f. figure 2-2) which lead to size reduction of particles in a combustor. The first two effects are caused by devolatilization of the fuel particles which yield coarse and fine char. The coarse char can also break down to fine char particles. By the burnout of the char finally pure ash is produced which is elutriated from the system or withdrawn as bottom ash. During the residence time of the ash it is undergoing attrition. The effects leading to the ash directly after burnout of the char are happening in a very short period of time. It is in the order of magnitude of some minutes. The result is then the so-called primary ash with its particle size distribution named as "primary ash particle size distribution" (PAPSD). The ash stays depending on its particles size for several hours, days or weeks in the system. During this period the particles are shrinking due to pure abrasion. Considering the time scale of these effects the abrasion effect of the ash is a very dominant one.

CAMMAROTA et al. (2001) developed therefore a method to determine the PAPSD of fuel particles under fluidizing conditions. They put some fuel particles in a metal basket, which is placed in a hot bubbling fluidized bed of quartz sand. After burnout the fragments of the particles are removed with the basket and their size is analyzed. The mesh width of the basket has to be big enough to allow the fluidized bed to penetrate the fuel particles. Only particle fragments can be collected which are bigger than the mesh. This method is a single particle experiment and only a small amount of ash is produced. Estimation of their attrition behavior is very difficult, as long as many experimental set-ups are requiring amounts of up to some kilogram.

Even all the work originated from or dealt with different systems of solids the attrition phenomena were quite similar and had common effects and mechanisms.

Except the degradation in the combustor attrition is always referred to pure abrasion. For combustion the attrition defined as pure abrasion is referred to ash particles after complete burnout defined as PAPSD (CAMMAROTA et al., 2002).

If a material tends to be easy attritable or not depends on properties of the particle which are material related, the origin of the particle, the fate of the particle and on the operating conditions where the particle is exposed to mechanical stress. These influences on the attrition of the solids will be discussed in the following.

Material Properties

Inner particle structure

The inner particle structure is one of the most important properties of a particle regarding its attrition behavior. The structure is decisive for the size and amount of fines produced and depends strongly on the origin of the particle. FCC catalyst can have a crystallized, amorphous or agglomerate like structure, respectively. The structure of the catalyst can be influenced by its way of production. On the other hand natural arisen particles like fossil fuels and their mineral matters which are left as ash after combustion have a natural inner structure which cannot be influenced and can differ strongly.

Particle Size

The particle size is of primary interest with respect to particle breakage, because the breakage probability strongly depends on the presence of microcracks or imperfections. Smaller particles are therefore more difficult to break than larger ones, mainly because they tend to contain fewer faults. The mechanisms of breakage will not be further discussed in the present chapter, but a comprehensive survey is given by the BRITISH MATERIALS HANDLING BOARD (1987).

On the other hand regarding attrition, which is a process, which takes place at the particle surface, small particles have a larger volume specific surface area than big ones. Therefore it is often discussed if attrition rates should be referred to the mass of particles or to its surface. RANGELOVA et al. (2004) compared model calculations using a mass related attrition coefficient (RANGELOVA et al., 2002) with calculations using a surface related one (RANGELOVA et al., 2004). For particles with long residence times they found significant differences in the particle size decay. Unfortunately no experimental data was available to proof which model variation would describe the reality best.

RANGELOVA et al. (2002 & 2004) assumed the mass or surface related attrition rate to be independent of the size of the individual particle whereas XI (1993) reported the jet-induced production of fines to be proportional to the particle size. By investigation regarding the influence of particle size on the attrition in cyclones REPPENHAGEN & WERTHER (2000) found the same relationship between mass related attrition rate and surface mean diameter. In analogy to these findings REPPENHAGEN & WERTHER (2001) assumed the particle size dependent attrition coefficient to be also proportional to the particle size for catalyst particles.

CAMMAROTA et al. (2001) stated the attrition rate of ash particles to be proportional to the reciprocal value of particle size d_p when a surface related attrition constant is used. Dividing their correlation for attrition in a bubbling fluidized bed with the mass specific surface area of a particle yields the production of fines per unit mass. Then the mass specific attrition rate becomes to be proportional to the particle size. This corresponds to the assumption of REPPENHAGEN & WERTHER (2001).

Particle Size Distribution

In the previous section the influence of the size of the individual particle is discussed without consideration of the whole particle size distribution of the bulk whether it is broad or narrow or has a high or low amount of fines or coarse particles, respectively. This might have an additional effect since FORSYTHE & HERTWIG (1949) already observed a reduction of the degradation of FCC-catalysts in jet attrition tests due to the presence of fines. They supposed some kind of cushioning effect, which limits the force of collision impact and thus limits the degradation of the coarse particles. This effect can now be explained with the foregoing findings. With more fines in the material the mass related attrition would decrease due to a lower mass specific amount of attritable coarse particles.

Particle Shape and Surface Structure

The particle shape is a relevant parameter, because irregular and angular particles are inclined to have their corners knocked off in collisions and thus become rounder and naturally smaller with time. A macroscopically smooth surface is therefore less prone to breakage, but it may still undergo abrasion. With respect to the latter the microscopic surface structure is of interest. Surface asperities may be chipped off and lead to abrasion.

Fate of the particles

The fate of the particles will be put here a closer look as it is in the focus of this work. The stress which particles have experienced in their "life" has a strong influence on their resistance to mechanical stress. It has to be distinguished between experienced stress before applied into a fluidized bed as long as this not within the focus of this work, but has an influence and will be explained short. The focus is set on the fate of the particles during exposition to mechanical stress within a fluidized bed system until the particles are leaving the system again.

Fate of particles within the fluidized bed system

The particle shape and surface structure is not constant with time. These properties are strongly dependent on the fate of the particles. Particles, which have undergone mechanical stress causing abrasion, have usually a much smoother surface than for example catalyst fresh from production or fresh generated ash, respectively. This is documented in literature for a lot of different materials. REPPENHAGEN & WERTHER (2000) documented the change of catalyst surface during attrition experiments regarding the attrition in cyclones by SEM analysis. RANGELOVA et al. (2004) made the same observation for mustard grains coated with lime. Therefore the fate of the particles has to be taken into account when modeling time dependencies of the particle size distribution changed by attrition.

The observations show a high attrition rate at the beginning of the experiment, which decreases with time to a constant attrition rate when the particles are rounded to a natural roughness. The decrease of the attrition rate shows for all materials the same general shape, which was first mathematically described for catalyst particles by GWYN (1969). In a batch fluidized bed process the elutriated mass m_{loss} with time t based on the initial bed mass $m_{bed,0}$ is

$$\frac{m_{loss}}{m_{bed,0}} = K_a \cdot t^b \qquad (2\text{-}1)$$

GWYN (1969) found the exponent b to be independent of particle size distribution whereas K_a is dependent on particle size for the above-mentioned reasons. PIS et al. (1991) and DESSALCES et al. (1994) suggested more complex correlations for the same observations. However, the time-dependence of attrition of catalyst particles is of secondary interest since in FCC processes the majority of the bed material in steady-state operation is usually already rounded of due to the long residence time.

For processes where continuously fresh material is fed, e.g. fluidized bed combustors with a continuous feed of ash generated from the fuel, sorbents and fresh bed material, the effect of "aging" is very important. DI BENEDETTO & SALATINO (1998) followed therefore the strategy that not the attrition coefficient for the steady-state attrition is taken for model calculations but the attrition rate is averaged over the first half hour of an experiment. The modeling is then made with this averaged attrition coefficient. This assumes that the average residence time of the particles in the fluidized bed is in the order of magnitude of half an hour.

As long as all investigations were dealing with the attrition in a bubbling fluidized bed without recirculation of solids there is a lot of knowledge missing how the attrition behavior regarding the fate of particles will be in a system with recirculation. With recirculation of solids particles are not only stressed in the fluidized bed itself but also in the cyclone for recirculation. The fact that not all particles are entrained from the bed in a time interval and that small particles are more often recirculated per unit time than big ones it is easy imagined that the influence of the cyclone on the state of a particle is not negligible. Even when all particles stayed the same time in the entire system they would have experienced different fates.

Fate of particles before fed into a fluidized bed system (pretreatment)

When the time or state dependence of attrition is discussed and applied for modeling the pretreatment and preparation of the particles have to be considered. When the time-dependence of a fresh material is determined in experiment the results can only be used for calculations, which starts also at the same point of state. For different plants and applications the solids are treated in a different manner on their way from production to application. In one application they could be conveyed to the usage by belt conveyers with nearly no mechanical stress and in another one by pneumatic transport. Under this circumstance the solids will enter the system of interest in one case nearly fresh and in the other case already pre-stressed, respectively.

Process Conditions

The process conditions have a strong influence on the attrition of particles by generating the stress on the individual particles. In general, the stress leading to attrition of a given bulk material may be a mechanical one due to compression, impact or shear, a thermal one owing to evaporation of moisture or temperature shock, or a chemical one by molecular volume change or partial conversion of the solid into the gas phase.

Gas and Solids Velocity

In a fluidized bed system the gas velocity is one of the most important parameter of the operating conditions and is directly decisive for the solids velocity. The solids velocity is the dominant factor in generating the mechanical stress by interparticle collisions or by particle-wall impacts (BRITISH MATERIALS HANDLING BOARD, 1987). Therefore the gas velocity is always in the focus when the dependencies of attrition are investigated.

Most investigations were made for the bubble-induced attrition. WERTHER & XI (1993) refer the attrition rate to be dependent on $(u\text{-}u_{mf})$ with a critical offset for FCC-catalyst. REPPENHAGEN (2000) improved the correlation of the attrition rate to be proportional to $(u\text{-}u_{mf})^3$. For fuel, ash and limestone DI BENEDETTO & SALATINO (1998) found the attrition rate to be proportional to $(u\text{-}u_{mf})$ and often assumes u_{mf} to be small compared with u and can therefore be neglected.

For the jet-induced attrition rate WERTHER & XI (1993) found the production of fines to depend on u^3 and in cyclones REPPENHAGEN & WERTHER (2000) determined the attrition rate to be proportional to u_{in}^2.

Wall-Hardness

One can assume that the particle degradation increases with the hardness of the vessel wall. This effect will increase with increasing ratio of particle-to-vessel diameter and may thus be only relevant inside a cyclone inlet duct, or in the vicinity of bed inserts.

Temperature, Pressure and Reaction Atmosphere in the System

Temperature, pressure and chemical reactions can have an influence on the attrition behavior of a particle due to changes in material properties such as strength, hardness and elasticity. As long as the conditions are kept constant these effects are negligible with respect to catalyst attrition. XI (1993) could not find an influence of the temperature level in the range of 20 to 400 °C for jet-induced attrition when the temperature of the jet is the same than of the surrounding fluidized bed. This is changed when the temperature of the jet is different to that of the bed. In the case of temperature differences the production of fines is proportional to $(T_b \; / \; T_j)^3$ when bed temperature T_b is smaller than the jet temperature T_j and proportional to the reciprocal value $(T_j \; / \; T_b)^3$ for the other way round (XI, 1993). In this case attrition might be a result by thermal shock as separately investigated by WHITCOMBE et al. (2003).

For attrition in combustors an additional influence is given by temperature and reaction atmosphere as defined by oxygen concentration which will influence the formation of the ash and therefore is decisive for inner structure, surface structure, hardness, etc. By the temperature of the combusting particle sintering effects might occur and with high temperatures ash slag might be formed which will yield to ash particles with different properties then ash generated at lower temperatures.

Moreover, both temperature and pressure can have a strong effect on the gas density, which affects the fluidization state and with it the particle motion and the stress the particles are subjected to.

2.2.2. Separation effects

Besides the change of the size of the individual particles by attrition and fragmentation the local particle size distribution in a fluidized bed system with recirculation of solids is influenced by the separation effects in the single components of the system, namely fluidized bed and cyclones.

Separation of solids in cyclones

A dominant role in the adjustment of the steady-state particle size distribution plays the separation performance of the cyclone. For FCC-reactors, which are often operated as a bubbling fluidized bed, the cyclone is decisive for the size of particles, which can be kept in the system. The segregation within the system is usually low due to the narrow particle size distribution. The system contains mostly particles, which can be entrained to the cyclone, and the particles are nearly homogeneous mixed in the bed. By the separation in the cyclone the fraction of fine particles is defined and the size of bubbles is strongly influenced which is of importance for the conversion of educts (WERTHER & HARTGE, 2004). They investigated with model calculations the influence of different cyclone configurations on the adjustment of particle size distributions and the consequences. A high separation efficiency of the cyclone does not always lead to the best performance with respect to minimize particle loss when attrition is considered besides the separation. REPPENHAGEN et al. (2000) combined the attrition in cyclones with the separation and determine then an optimum operating condition of a given cyclone to minimize material loss.

In circulating fluidized beds with a rather broad particle size distribution the segregation effects are obvious. HERBERTZ et al. (1989) investigated the particle size distributions at different locations in a circulating fluidized bed combustor and found strong differences due to segregation within the combustion chamber and separation in the recirculating cyclone.

As mentioned in 2.1 LALAK et al. (2003) could proof the big influence of the cyclone. As a result the operation of the cyclone has to be described very precise in the prediction of the particle size distribution. Some models are available for calculation of separation in cyclones. Ufortunately most of them are dealing with very lean gas solids suspensions (LEITH & LICHT, 1972; MOTHES & LÖFFLER, 1984). When the solids loading of the gas stream is above the so-called critical solids loading a second separation effect occurs besides the separation by the centrifugal force field in the inner vortex of the cyclone. The vortex flow field of a cyclone seems to have only a limited carrying capacity for the solids. When the concentration of solids in the incoming gas is higher than a certain value the amount of solids above the critical solids loading settles spontaneously direct at the cyclone inlet. The settled solids flow in form of a strand down the cyclone wall towards the lower outlet. The critical solids load depends on the gas properties, volume flow rate and the solids particle size distribution, respectively. In literature there is only one approach available for the description of this effect by MUSCHELKNAUTZ (1970) and KRAMBROCK (1971), which is further improved to the actual state in TREFZ & MUSCHELKNAUTZ (1993).

More recent investigations regarding separation in cyclones are dealing with more sophisticated geometries of the cyclone as position of vortex finder (MUSCHELKNAUTZ & MUSCHELKNAUTZ, 1996 & 1999), the inlet geometry (HUGI & REH, 1998) or the influence of the down-comer configuration below the cyclone underflow (OBERMAIR & STAUDINGER, 2001). All these investigations yield further knowledge how the separation of cyclones can be improved but have not yet been considered in modeling. The design procedure of TREFZ & MUSCHELKNAUTZ (1993) can therefore only be strictly taken for standardized cyclones which were the basis of the experimental investigations but will be also applied for other geometries with some uncertainties.

Entrainment of solids from the fluidized bed

The entrainment of solids from a fluidized bed is the second separation effect occurring in fluidized bed systems and of high interest since it is decisive for the design and operation conditions of the cyclone for the recirculation of the solids. By the entrainment the mass flux of solids and the particle size distribution of the inflowing solids are defined.

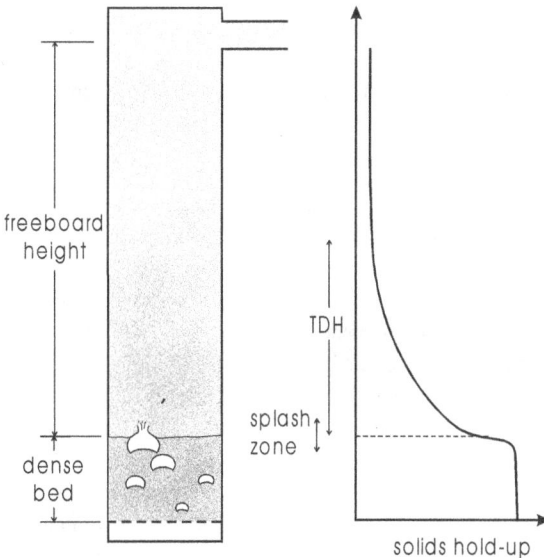

Figure 2-3: Zones in a fluidized bed vessel

The fluidized bed is usually divided in different vertical zones (cf. figure 2-3). At the bottom there is the dense fluidized bed also called the bottom zone or bed, above which the freeboard or upper dilute zone is located. The transient region in between is denoted as the splash zone where rising bubbles from the bottom bed explode and cause a more fuzzy transition between bottom bed and upper dilute zone. From this region the solids hold-up gradually decays further until it becomes constant or at least nearly constant. The distance between this point, where the solids concentration becomes nearly constant, and the surface of the fluidized bed is called the 'Transport Disengaging Height" (TDH).

Two effects cause the entrainment of particles from the fluidized bed and since small particles tend to a greater extent to be entrained than big ones also cause the separation. One effect is the ejection of particles into the freeboard by exploding bubbles (LEVY et al., 1983). After the ejection into the freeboard with a certain starting velocity the particles will be decelerate and either fall back to the bottom bed or be transported by the up-flowing gas dependent on their terminal velocity being smaller or greater then the gas velocity. Here the assumption is usually made that the particles do not interact with each other (ZENZ & WEIL, 1958, DO et al., 1972).

Some operating parameters have a strong influence on the entrainment of solids. Since the influence of the superficial gas velocity of the fluidizing gas is obvious some other parameters should be discussed shortly. The size of particles is decisive in combination with the particle density for its terminal velocity and therefore transportability by the up-flowing gas. For fine particles with terminal velocity w_s below the gas velocity in the freeboard the elutriation rate increases with decreasing particles size (TASIRIN & GELDART, 1998). The particle size distribution has an influence on the amount of particles of a certain size which are ejected to the freeboard. It is usually assumed that particles are homogeneous mixed in the bottom bed and therefore the exploding bubbles are ejecting solids with representative particle size distribution.

Furthermore the geometry of the freeboard has an influence on the transport of the particles. Expansion or narrowing, respectively, is changing gas velocity with influence on carrying capacity of the gas. The height of the freeboard is of big interest. Close to the bed surface the carryover from the fluidized bed is decreasing significantly with increasing height. The decay in the entrainment becomes nearly constant, when the Transport Disengaging Height (TDH) is exceeded (cf. figure 2-3). Similar to the entrainment flux the solids volume concentration in the freeboard also decreases with increasing distance from the bed surface and becomes nearly constant above the TDH. A lot of studies were carried out in laboratory scale riser with small diameters which raises the question of the influence of diameter. Only few studies of the influence of scale on the entrainment are available in the literature. LEWIS et al. (1962) made a study on entrainment with units of 0.019 to 0.146 m diameter. As a result they stated that entrainment is independent of the size for units larger than 0.1 m in diameter. This result is supported by findings of Colakyan & Levenspiel (1984) and Tasirin & Geldart (1998b).

Further influencing parameters and literature overview is given in Kunii & Levenspiel (1991) and more recently by Werther & Hartge (2003).

Finally the remark should be made that unfortunately the investigations and results of separation effects of cyclones and fluidized beds were made with systems containing solids of one material and are not yet been verified or extended for mixtures of solids with different densities, respectively.

2.3. Modeling of the particle size distribution by population balances

For the mathematical description of a given fluidized bed system the above mentioned effects and their single models have to be combined in one model for the whole system. This is usually done with population balances for every single size class.

How to balance and which property has to be balanced is recently in the focus of discussion. The particle size distribution of solids can either be described in terms of masses or numbers in a certain size interval. As long as not every individual particle can be characterized in term of shape and size some approximations have to be made. Most effects happening in a process are happening to the individual particle and therefore it is seductive to describe the effects on the basis of numbers. But on the other hand the model results are validated with measurements where the total mass or volume of particles within one size interval is determined. Then the number based distribution has to be transformed to a mass based distribution and vice versa. By these transformations big errors might occur and the comparability is questionable. When the number balances are closed the mass balances are usually not and the other round but the closure of the mass balance is the one which can be validated by experiments easily. Therefore the modeling should base on the primary measured variable (Scarlett, 2002).

3. Theory

3.1. Strategy for modeling the particle size distribution in a fluidized bed system with recirculation of solids

A fluidized bed system can consist of a combination of fluidized beds, cyclones and return lines. A simple combination is presented in figure 3-1. When a flexible model is desired for the description of effects concerning the particle size distribution (PSD) of the solids in the system it is wise to subdivide the model into modules. Each module describes one single apparatus of the system and provides an interface which delivers the incoming streams for the next module. The modules defined here are as mentioned fluidized beds in two different flow regimes, namely bubbling fluidization with moderate gas velocities and fast fluidization used for circulating fluidized bed systems. Additionally, one module is needed for the effects occurring in cyclones and one module for the description of return lines.

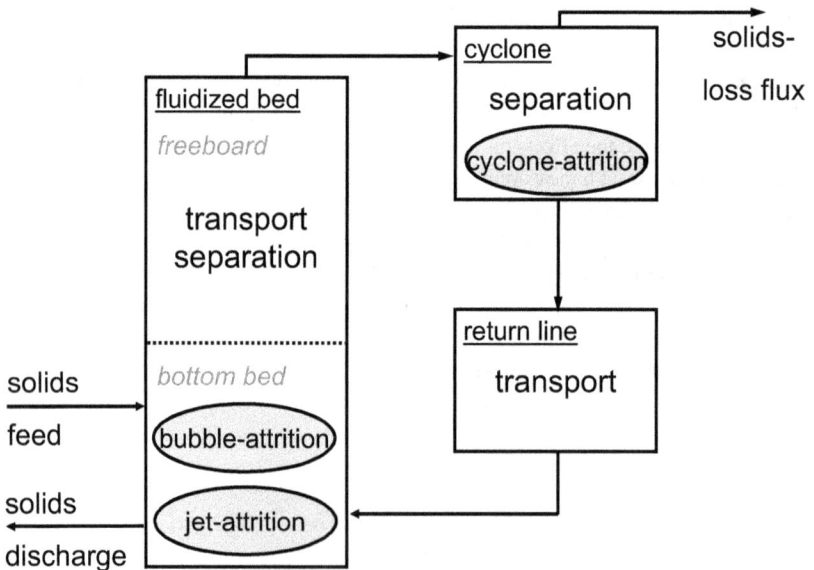

Figure 3-1: Example of the model layout of a fluidized bed system with recirculation of solids.

The PSD relevant effects considered in the modules can be separated into two groups, namely attrition which is changing the size of particles and separation which influences the PSD of the particles. Additionally the solids feed has an influence when particles with differing PSD from the bed inventory PSD are fed. Solids discharge is changing only the total mass of bed inventory.

The model calculations for shrinkage due to attrition and separation effects within the modules fluidized bed and cyclone are explained in the following sections. The attrition effects are described separately from the separation. It is in general the same for each module. The return line is modeled as a stand pipe with plug flow where solids can be accumulated and have therefore an influence on time dependent effects. It does not need to be explained in detail. Besides the return line module only the fluidized bed module is capable of accumulating solids. In the cyclone it is assumed that solids accumulation is negligible.

The dynamic calculation of such a system is realized in such a way that the modules are initialized with a defined status, e.g. mass of inventory, particle size distribution, stress history distribution, kind of solids and model specific parameters, etc. From there on the modules are calculated sequentially with a prefixed time step length Δt. The result of one module is an outflowing stream which is given as the inflowing stream into the next module in the actual time step and so on. When the time steps are small enough a good approximation to time dependent behavior can be reached.

When attrition and change in experienced stress is occurring in one of the modules the consequences, change in particle size and stress related state, are calculated first and then the separation and transport effects are simulated. This is valid in the fluidized bed for the whole inventory and in the cyclone for the incoming mass portion of solids within this time step. The modeling of shrinkage and time-dependence of attrition of particles is explained for both the modules in the following two sections.

The described model will be later applied exemplarily to two different systems of fluidized beds. One is a bubbling fluidized bed with FCC catalyst as single component system operated batch-wise.

The other one is a circulating fluidized bed combustor (CFBC) which will operate continuously and is a multicomponent system containing quartz sand, sorbent particles and ashes of two different fuels. Since the developed model only considers the development of the PSD influenced by mechanical mechanisms the conversion of fuel to ash by combustion is not implemented. Therefore the strategy here will be to assume that the incoming stream of fuel is replaced by a fictitious stream of ash. With the knowledge of the ash content of the fuel the total mass flow of ash can easily be determined. But the particle size distribution of the fresh ash has to be put as an input parameter additionally. This distribution is called the primary ash size distribution (PAPSD) and is defined to be the ash particle size distribution directly after the complete burnout and primary fragmentation of loose agglomerates (Cammarota et al., 2002) (cf. 2.2.1).

3.2. Attrition in a fluidized bed system

3.2.1. Change of particle size distribution by attrition

The mass balance for one class i which ranges from x_{i-1} to x_i can be written as followed:

$$\frac{dm_i}{dt} = -\dot{m}_{att,i} - \dot{m}_{i,i-1} + \dot{m}_{i+1,i} \tag{3-1}$$

The loss of mass in the class i leaving the class as produced fines is $\dot{m}_{att,i}$. The production of fines leads to the shrinkage of particles which causes some particles to shrink from class i to class $i-1$. This mass moving across the boundary from class i to class $i-1$ is the mass flow rate $\dot{m}_{i,i-1}$. In figure 3-2 the change of mass due to attrition is visualized. In the drawing the masses are sketched as mass portions which are changed within a time step Δt.

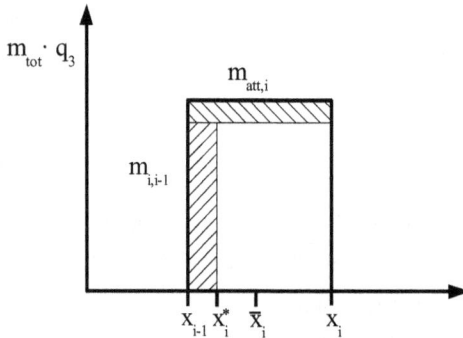

Figure 3-2: Change of mass in size class i by attrition of particles

With the assumption of an even distribution of mass within the size interval between x_{i-1} and x_i (which is plotted in figure 3-2) the mass of solids traveling from class i to class $i-1$ is calculated by

$$m_{i,i-1} = (m_i - m_{att,i}) \cdot \frac{x_i^* - x_{i-1}}{x_i - x_{i-1}} \tag{3-2}$$

x_i^* is the diameter of the biggest particle which will shrink enough to move from class i to class $i-1$ within time interval Δt. The mass of the particle with size x_i^* can be determined by solving the mass balance at a single particle (index p)

$$m_{p,i}^* = m_{p,i-1} + r_i \cdot \Delta t \cdot m_{p,i}^* . \tag{3-3}$$

r_i is a mass related attrition rate for the particles in class i with approximation that the attrition rate is equal for all particles within class i and defined as follows.

$$r_i = \frac{\dot{m}_{att,i}}{m_i} \qquad (3\text{-}4)$$

The attrition rate r_i is dependent on the source of attrition, operating conditions, solids properties and the state of the particles. The definition and calculation of r_i will be explained in the following section. Equation (3-3) can be written as

$$x_i^{*^3} = x_{i-1}^3 + r_i \cdot \Delta t \cdot x_i^{*^3} \qquad (3\text{-}5)$$

and x_i^* is then calculated by

$$x_i^* = \frac{x_{i-1}}{\sqrt[3]{1 - r_i \cdot \Delta t}} \qquad (3\text{-}6)$$

3.2.2. Time-dependence of attrition

The typical time dependent development of the production of fines in experimental investigations on attrition is sketched in figure 3-3. This behavior is observed for the jet-induced attrition, the bubble-induced attrition and also for attrition in cyclones. When an experiment is started with fresh particles, which have not undergone attrition before, the production rate of fines is high at the beginning, decreases and finally approaches a constant rate.

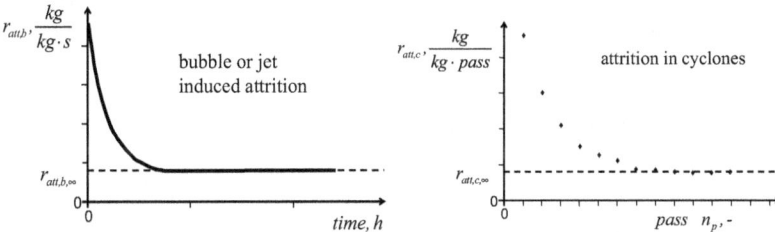

Figure 3-3: Qualitative development of attrition rate in bubbling fluidized beds with time

The dependence of the steady-state attrition rate r_∞ on operating condition and material properties was investigated and modeled by former scientists (WERTHER & XI, 1993; WERTHER & REPPENHAGEN, 1999; REPPENHAGEN & WERTHER, 2000 & 2001). r_∞ is the production rate of fines $\dot{m}_{att,\infty}$ related to the mass of solids exposed to stress. $\dot{m}_{att,\infty}$ under steady-state conditions can be described for all sources by the following relationship.

$$\dot{m}_{att,\infty}^* = C^* \cdot d_p \cdot \xi^* \qquad \left| \xi^* = \xi_b, \xi_j, \xi_c; \quad C^* = C_b, C_j, C_c \right. \tag{3-7}$$

Here the attrition at steady-state conditions is calculated in dependence of a solids related attrition coefficient C^*, a characteristic mean particle diameter (surface mean diameter) d_p and a parameter ξ^* which is taking the operating conditions at the source into account with

$$\xi_b = m_b \cdot (u - u_{mf})^3 \tag{3-8}$$

$$\xi_j = \rho_f \cdot d_{or}^2 \cdot u_{or}^3 \cdot n_{or} \tag{3-9}$$

$$\xi_c = \dot{m}_{c,in} \cdot u_{in}^2 / \sqrt{\mu_{in}} \tag{3-10}$$

It was found that the production of fines is proportional to the surface mean diameter of the particle size distribution which is calculated as follows for a discretisized particle size distribution:

$$d_p = M_{1,2} \cong \sum \bar{x}_i \cdot \Delta Q_{2,i} \tag{3-11}$$

By introducing eqn. (3-11) in eqn. (3-7) the total production mass rate of fines is

$$\dot{m}_{att,\infty} = \sum \dot{m}_{att,\infty,i} = C^* \cdot \xi^* \cdot \sum \bar{x}_i \cdot \Delta Q_{2,i} \tag{3-12}$$

where the production rate of fines within one class i can be deduced to be

$$\dot{m}_{att,\infty,i} = C^* \cdot \xi^* \cdot \bar{x}_i \cdot \Delta Q_{2,i} \tag{3-13}$$

When the production of fines is expressed as a dimensionless attrition rate r_i the produced mass of fines $\dot{m}_{att,\infty,i}$ is related to the mass m_i of solids which are attrited. For one class i the attrition rate r_i is then

$$r_{\infty,i} = \frac{\dot{m}_{att,\infty,i}}{m_i} = \frac{C^* \cdot \xi^*}{m_{tot}} \cdot \bar{x}_i \cdot \frac{\Delta Q_{2,i}}{\Delta Q_{3,i}} \tag{3-14}$$

For the mass related attrition rate $r_{\infty,tot}$ of the whole bulk this means

$$r_{tot,\infty} = \frac{\sum r_{i,\infty} \cdot m_i}{m_{tot}} = \sum r_{i,\infty} \cdot \Delta Q_{3,i} \tag{3-15}$$

As is seen from experiments the attrition rate is not only depending on operating conditions and material properties but also on the state of the particles. The momentary state of the particle and its history plays an important role for the attrition rate. When the surface of the particles is not yet smoothed enough they will produce much more fines than under the steady-state conditions. For the description of the development from a fresh particle with a lot of edges to a rounded smooth one a relationship has to be found.

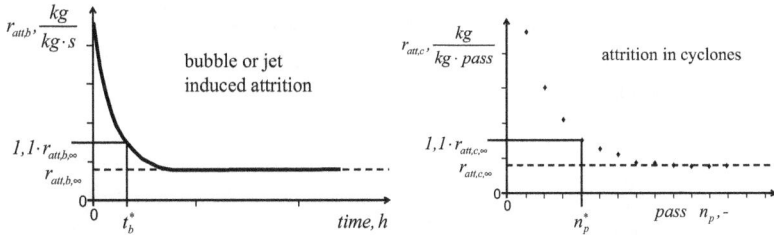

Figure 3-4: Definition of the characteristic parameters t^* and n_c^* for introduction of the dimensionless stress history parameter ϑ

For convenience, in the current work the time t^* needed to reach a production rate of fines which is 10% above the final steady-state attrition rate has been chosen to characterize the time dependent behavior of the attrition. This definition is visualized in figure 3-4 at the example of attrition induced by bubbles. The same procedure is applied for the attrition due to jets. In analogy it is done for attrition in cyclones with the difference that the attrition in cyclones is not time-dependent, but depending on the number of passes n_c through the cyclone. Therefore the characteristic value is called n_c^*. Based on the characteristic values t^* or n_c^* a stress history parameter ϑ is defined such, that ϑ is equal to 1 for $t = t^*$ or $n_c = n_c^*$, respectively.

$$\vartheta_b = t / t_b^* \qquad \text{for in-bed attrition} \qquad (3\text{-}16)$$

$$\vartheta_c = n_c / n_c^* \qquad \text{for attrition in cyclones} \qquad (3\text{-}17)$$

$$\vartheta_j = t / t_j^* \qquad \text{for jet-induced attrition} \qquad (3\text{-}18)$$

In a fluidized bed system with solids recirculation particles with different sizes have different residence times in the different regions of the system. This means, that the stress histories for the different mechanisms are coupled, e.g. stress which the particle experiences in the cyclone will also influence the subsequent attrition rate due to bubbles. For simplicity it is assumed in the following that the stresses which the particles experience due to the different mechanisms can be added. Then a uniform description of stress for all three attrition mechanisms is needed. The attrition rates in the different sources can then be described by

$$r_i(\vartheta) = f(\vartheta) \cdot r_{\infty,i} \tag{3-19}$$

with $r_{\infty,i}$ calculated according to eqn. (3-14). The influence of the experienced stress history is here introduced by the factor $f(\vartheta)$.

On the basis of experiments a mathematical relationship for $f(\vartheta)$ has to be found. As will be shown later the experiments conducted during this work and for the solids used can be described by the following relationship.

$$f(\vartheta) = \begin{cases} 1.1 \cdot \vartheta^b & \vartheta \le (1/1.1)^{1/b} \\ 1 & \vartheta > (1/1.1)^{1/b} \end{cases} \tag{3-20}$$

With the introduction of ϑ as the stress history parameter a second dimension for the solids property besides their size is added and has to be considered in calculations. In order to take account of the different stress histories of particles the mass of particles in each size class is assumed to be distributed over several stress history classes with ϑ_j as the upper bound of class j. Particles are then transferred according to their individual stress from stress history classes with small ϑ to classes with larger ϑ. The mass transferred within a time step Δt in the bed or in the jet or one pass through the cyclone from class j to class $j+1$ can be determined by

$$m_{j,j+1} = m_j \cdot \frac{\Delta \vartheta}{\vartheta_{j+1} - \vartheta_j} \tag{3-21}$$

Here it is assumed that particles are evenly distributed in the interval of ϑ_j to ϑ_{j+1} and $\Delta \vartheta$ is describing the aging of material during one time step Δt or one pass through the cyclone, respectively.

$$\Delta \vartheta_b = \Delta t / t_b^* \quad \text{for in-bed attrition} \tag{3-22}$$

$$\Delta \vartheta_c = 1/n_p^* \quad \text{for attrition in cyclones} \tag{3-23}$$

$$\Delta \vartheta_j = \Delta t / t_j^* \quad \text{for jet-induced attrition} \tag{3-24}$$

The values for t_b^* and n_c^* are directly determined from experiments. The same is done for t_j^* with an important difference. WERTHER & XI (1993) concluded from their results that the absolute production rate of fines for a single jet is independent of the mass of the fluidized bed surrounding the jet as long as the jet is completely covered with solids, which means the bed is high enough. From this result it can be concluded that a jet stresses only a fixed amount of solids in a time step. With this knowledge the value of t_j^* has to be scaled from the experimental conditions to the conditions which have to be modeled. A single jet can only transfer a fixed amount of particles from one stress class j to the next class $j+1$. t_j^* for calculation is then obtained from $t_j^*{}_{exp}$ which was measured in experiments as follows:

$$t_j^* = t_{j,exp.}^* \cdot \frac{m_b}{m_{b,exp.}} \cdot \frac{n_{or,exp.}}{n_{or}} \tag{3-25}$$

Here $m_{b,exp.}$ and $n_{or,exp.}$ are the mass of bed material and number of jets in the experimental determination of $t_{j,exp.}^*$ and m_b and n_{or} for the calculation conditions, respectively.

3.3. Separation effects in a fluidized bed system

3.3.1. Separation in cyclones

For recirculation of solids in a fluidized bed system cyclones are commonly used. By passing the cyclone the particles are not only exposed to mechanical stress and attrited as shown before but they are also separated by size. To be very precise the particles are separated in the centrifugal flow field of the cyclone by their different terminal velocities in the flow field which is a consequence of their size and density. Common models for the calculation of the separation in cyclones are based on the assumption of having particles of the same density. But there are a lot of processes, where particles from different materials are involved, e.g. combustion and chemical reactors, where the solids are converted from one compound to another one.

To consider the differences in density a cyclone model is needed which is based on a terminal velocity distribution of the particles. The model presented here is based on the model suggested by TREFZ & MUSCHELKNAUTZ (1993) which is commonly used in Germany in industrial application and design procedures. To make it applicable for the mentioned purpose some modifications where made which will be explained in the following section.

The calculation of the flow field, e.g. velocities, is taken without any modifications. The calculation of the separation has to be modified to consider differences in densities of the particles. Two effects cause the separation of the solids from the gas, one is the classification in the vortex due to the centrifugal forces, the other one is due to the limited solids carrying capacity of a gas flow. This latter separation takes place in the vicinity of the inlet.

For loadings μ higher than a critical loading μ_G a solid mass corresponding to $(\mu - \mu_G)$ is separated from the flow in the form of strands and only the mass fraction limited by μ_G enters the vortex.

41

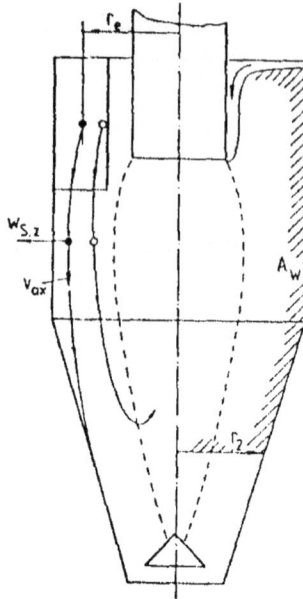

Figure 3-5: Axial and radial velocity components of a particle on mean streamline through a cyclone (from TREFZ & MUSCHELKNAUTZ**, 1993).**

The classification in the main flow field may be explained with the help of the model illustrated in figure 3-5. A centrifugal force Z acts on a particle, which is carried by the gas flow through the cyclone, and causes a radial settling velocity of $w_{s,z}$.

According to figure 3-5, this settling velocity is superimposed by an axial flow component v_{ax}. It is obvious that very coarse particles with a high settling velocity will reach the wall and the finer particles may be dragged to the inner vortex where they are classified according to centrifugal forces and an efficiency grade curve.

Whether a particle becomes separated at the wall or not, depends on the main volumetric flow rate $\dot{V} - \dot{V}_{sec}$ and the clarification area A_w. Assuming a mean secondary flow of 10 %, TREFZ & MUSCHELKNAUTZ (1993) obtain for the settling velocity $w_{s,50}$ of a particle, which is separated in the zone of downward flow with a probability of 50 %,

$$w_{s,50} = \frac{0.5 \cdot 0.9 \cdot \dot{V}}{A_w} \tag{3-26}$$

Assuming the validity of Stokes' law, a cut size d_e^* for this zone may be calculated as a function of gas viscosity η_f and difference in density $\Delta\rho$:

$$d_e^* = \sqrt{w_{s,50}\frac{18\eta_f}{\Delta\rho \cdot \overline{z}}} \qquad (3-27)$$

Based on d_e^* TREFZ & MUSCHELKNAUTZ (1993) give a relationship for the maximum total solids load in the inner vortex to be:

$$\mu_G = 0.025 \cdot \left(\frac{d_e^*}{d_{50}}\right) \cdot (10 \cdot \mu_{in})^k \qquad (3-28)$$

The exponent k is given by

$$k = \begin{cases} 0.81 & for & \mu_{in} < 2.2 \cdot 10^{-5} \\ 0.15 + 0.66 \cdot exp\left(-\left(\frac{\mu_{in}}{0.015}\right)^{0.6}\right) & for & 2.2 \cdot 10^{-5} \le \mu_{in} \le 0.1 \\ 0.15 & for & 0.1 < \mu_{in} \end{cases}$$

$$(3-29)$$

To consider the terminal velocity of the particles the equations have to be modified. Therefore a critical solid load is defined for each size class to be

$$\mu_{G,i}(w_{s,i}) = 0.025 \cdot \left(\sqrt{\frac{w_{s,50}}{w_{s,i}}}\right) \cdot (10 \cdot \mu_{in})^k \qquad (3-30)$$

This solid load would be the critical load in the inner vortex, when the whole incoming stream contains only particles of this size or terminal velocity, respectively. Considering the mass fraction $\Delta Q_{3,in,i}$ of class i the class-wise solid load in the inner vortex is then

$$\mu_i(w_{s,i}) = \begin{cases} \Delta Q_{3,in}(w_{s,i}) \cdot \mu_{G,i}(w_{s,i}) & for & \Delta Q_{3,in}(w_{s,i}) \cdot \mu_{G,i}(w_{s,i}) < \Delta Q_{3,in}(w_{s,i}) \cdot \mu_{in} \\ \Delta Q_{3,in}(w_{s,i}) \cdot \mu_{in} & for & \Delta Q_{3,in}(w_{s,i}) \cdot \mu_{G,i}(w_{s,i}) \ge \Delta Q_{3,in}(w_{s,i}) \cdot \mu_{in} \end{cases}$$

$$(3-31)$$

For the mass fraction in the inner vortex this means

$$\Delta Q_{3,i}(w_{s,i}) = \frac{\mu_i(w_{s,i})}{\sum \mu_i(w_{s,i})} \qquad (3-32)$$

The terminal velocity $w_{s,i}$ of class i has to be determined considering the centrifugal flow field with the centrifugal acceleration $\bar{z}_e$ (see below). The mean centrifugal acceleration $\bar{z}_e$, along the streamline depends on the flow field. It is calculated from the tangential velocities $v_{\varphi e}$

$$v_{\varphi e} = \frac{v_{\varphi a} \cdot r_a / r_e}{1 + \frac{\lambda_s}{2} \cdot \frac{A_{el}}{0.9 \cdot \dot{V}} \cdot v_{\varphi a} \cdot \sqrt{r_a / r_e}} \tag{3-33}$$

with

$$A_{el} = \frac{\left(2 \pi r_a h_e\right)}{2} \tag{3-34}$$

and $v_{\varphi 2}$

$$v_{\varphi 2} = \frac{v_{\varphi a} \cdot r_a / r_2}{1 + \frac{\lambda_s}{2} \cdot \frac{A_w}{0.9 \cdot \dot{V}} \cdot v_{\varphi a} \cdot \sqrt{r_a / r_2}} \tag{3-35}$$

Assuming that only a small part of the volumetric flow rate $\dot{V} - \dot{V}_{sec}$ reaches the bottom of the cone, the area A_w is limited by the radius r_2, which bisects the height of the cone.

This gives for the mean centrifugal acceleration $\bar{z}_e$

$$\bar{z}_e = \frac{v_{\varphi e} \cdot v_{\varphi 2}}{\sqrt{r_e \cdot r_2}} \tag{3-36}$$

The separation in the inner vortex is determined by the cut size d_p^*. The non ideal separation can be described by a separation efficiency function depending on the particle diameter d normalized by the cut size d_p^* given by TREFZ &MUSCHELKNAUTZ (1993)

$$\eta_F = \begin{cases} 0 \ \text{für} \ x/d_p^* \leq 1/D \\ 0.5 \left\{ 1 + cos\left[\pi \left(1 - \frac{log\left(x/d_p^*\right) + logD}{2 \cdot logD} \right) \right] \right\} \ for \ 1/D < x/d_p^* < D \\ 1 \ \text{für} \ x/d_p^* \geq D \end{cases} \tag{3-37}$$

The parameter D depends on the cyclone geometry and is in the range between 2 and 4 and can be estimated for a typical cyclone as $D = 3$. The cut size of the inner vortex is the size of particles with a settling velocity of being equal to the radial velocity on the radius of the vortex finder

$$v_{r,vf} = \frac{0.9 \cdot \dot{V}}{2\pi r_{vf} h_{vf}} \tag{3-38}$$

Following the argumentation for the calculation of separation in the inlet region above based on terminal velocities the fractional separation within the inner vortex has to be then

$$\eta_{V,i}(w_{s,i}) = \begin{cases} 0 \; für \; \sqrt{w_{s,i}/v_{r,vf}} \leq 1/D \\ 0.5 \left\{ 1 + cos\left[\pi \left(1 - \frac{log\left(\sqrt{w_{s,i}/v_{r,vf}}\right) + log D}{2 \cdot log D} \right) \right] \right\} \; for \; 1/D < \sqrt{w_{s,i}/v_{r,vf}} < D \\ 1 \; für \; \sqrt{w_{s,i}/v_{r,vf}} \geq D \end{cases} \tag{3-39}$$

For calculation of terminal velocities $w_{s,i}$ under flow conditions in the inner vortex the centrifugal acceleration is calculated with vortex finder radius and tangential velocity $v_{\varphi,vf}$ given by TREFZ & MUSCHELKNAUTZ (1993).

$$z_{vf} = \frac{v_{\varphi,vf}^2}{r_{vf}} \tag{3-40}$$

The total fractional separation efficiency for class i can be calculated as a serial of separation effects and the mass of incoming solid $m_{c,in,i}$ of class i is then distributed between underflow $m_{c,u,i}$ and overflow $m_{c,o,i}$ as follows:

$$m_{c,u,i} = \left[1 - (1 - \eta_{V,i}) \cdot \frac{\mu_{G,i}}{\mu_{in,i}} \right] \cdot m_{c,in,i} \tag{3-41}$$

$$m_{c,o,i} = \left[(1 - \eta_{V,i}) \cdot \frac{\mu_{G,i}}{\mu_{in,i}} \right] \cdot m_{c,in,i} \tag{3-42}$$

3.3.2. Separation by entrainment from the fluidized bed

With regard to the description of the fluidized bed a distinction is made between two different kinds of fluidizing regimes. In a bubbling fluidized bed with moderate fluidizing velocities nearly the whole amount of the bed inventory is accumulated in the fluidized bed and negligible masses of particles are in the freeboard above the bed surface. Fast fluidization leads to a high accumulation of solids in the freeboard. Then the bed inventory has to be divided between the bottom bed and the upper dilute zone. These differences have to be considered in the modeling.

For both of the fluidizing regimes it is assumed that a multi-hole-distributor which has n_{or} orifices is installed.

<u>Bubbling fluidization</u>

In a bubbling fluidized bed the assumption of a homogeneous distribution of particles is made. The solids are entrained from the bed by bubbles exploding at the bed surface and when the freeboard is higher than the so-called transport disengaging height (TDH) the elutriated mass of solids is depending on the fluidizing gas velocity, the terminal velocity of particles and their mass fraction in the inventory (KUNII & LEVENSPIEL, 1991). The elutriated mass $m_{e,i}$ in the time step Δt of class i can be calculated by

$$m_{e,i} = k_{\infty,i} \cdot \Delta Q_{3,bed,i} \cdot \Delta t \cdot A_t \qquad (3\text{-}43)$$

$k_{\infty,i}$ is a particle related entrainment coefficient with the dimension of kg/m²/s. This coefficient is the mass flow of particles of size i which would occur when the whole inventory contains only particles of size i and is referred to the cross section of the vessel. It has then to be multiplied by the mass fraction of class i on the total inventory, the time step length and the cross sectional area of the vessel.

$k_{\infty,i}$ is a function of the terminal velocity of the particles and the fluidizing velocity. Several correlations for $k_{\infty,i}$ are given in the literature. Some overviews of the available correlations are can be found in KUNII & LEVENSPIEL (1991) and WERTHER & HARTGE (2003). In the calculations presented in this work the correlation of TASIRIN & GELDART (1998b) is applied.

$$k_{\infty,i}\left[\frac{kg}{m^2 s}\right] = \begin{cases} 23.7 \cdot \rho_f \cdot u^{2.5} \exp\left(-5.4\,\frac{w_{s,i}}{u}\right) & \text{for} \quad Re < 3000 \\[2mm] 14.5 \cdot \rho_f \cdot u^{2.5} \exp\left(-5.4\,\frac{w_{s,i}}{u}\right) & \text{for} \quad Re > 3000 \end{cases}$$

$$(3\text{-}44)$$

with $\quad Re = \dfrac{D_t \cdot u \cdot \rho_f}{\eta_f}$

The decision for one correlation depends on the system to be simulated. The correlations where gained in experimental work and are therefore only valid within the investigated ranges defined by kind and size of solids, size of fluidized bed and superficial fluidizing gas velocity.

Circulating fluidized bed

As mentioned above for circulating fluidized bed regimes there are two zones in the riser. The bottom bed in the lower part of the riser is characterized by high solids volume concentrations. The bottom zone is modeled as a bubbling fluidized bed according to WERTHER & WEIN (1994).

In figure 3-6 the general shape of the axial solids volume concentration profile is presented. The two zones of dense phase at the bottom and continuously decrease of solids volume concentration in the upper dilute zone can be differentiated obviously.

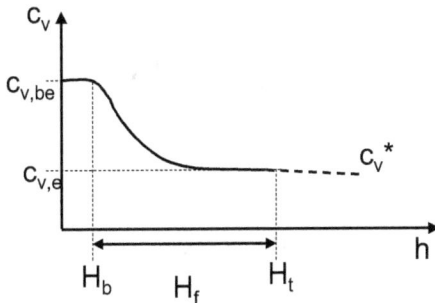

Figure 3-6: Axial solids volume concentration profile for fast fluidization

With the superficial fluidizing velocity u we obtain the volumetric flow $\dot{V}_{or}$ through a single opening of a nozzle of the distributor plate as

$$\dot{V}_{or} = \frac{u \cdot A_t}{n_{or}} \tag{3-45}$$

which according to DAVIDSON & HARRISON (1963) will lead to the formation of bubbles with an initial diameter of the volume equivalent sphere.

$$d_{v,0} = 1.3 \cdot \left(\frac{\dot{V}_0^2}{g} \right)^{0.2} \tag{3-46}$$

Since we are considering a multihole distributor these bubbles will be formed at the tip of the jets issuing from the individual holes.

Based on the pioneering work by CLIFT AND GRACE (1970, 1971) on the mechanism of bubble coalescence in gas fluidized beds WERTHER (1976) developed a model for bubble growth in freely bubbling fluidized beds which was later on extended to the consideration of the combined action of coalescence and splitting of the bubble by HILLIGARDT & WERTHER (1987). According to the latter work the growth of the bubble volume-equivalent sphere diameter d_V with height h is for Geldart A and B powders given by

$$\frac{d(d_v)}{dh} = \left(\frac{2 \cdot \varepsilon_b}{9 \cdot \pi}\right)^{1/3} - \frac{d_v}{3 \cdot \lambda \cdot u_b} \tag{3-47}$$

where the first term describes growth by coalescence and the second term accounts for bubble splitting. λ is the mean duration of life of a bubble with diameter d_V which according HILLIGARDT & WERTHER (1987) to can be calculated from

$$\lambda = 280 \cdot \frac{u_{mf}}{g} \tag{3-48}$$

The local bubble volume fraction ε_b is given by

$$\varepsilon_b = \frac{\dot{V}_b}{u_b} \tag{3-49}$$

where the visible bubble flow $\dot{V}_b$ may be approximated for a Geldart A powder by

$$\dot{V}_b \approx 0.8 \cdot \left(u - u_{mf}\right) \tag{3-50}$$

The local bubble rise velocity u_b is given by HILLIGARDT & WERTHER (1987)

$$u_b = \dot{V}_b + 0.71 \cdot \vartheta_v \cdot \sqrt{g \cdot d_v} \tag{3-51}$$

where for a Geldart A powder

$$\vartheta_v = \begin{cases} 1.18 & D_t < 0.05 \\ 3.2 \cdot D_t^{0.33} & 0.05 \leq D_t < 1 \\ 3.2 & D_t > 1 \end{cases} \tag{3-52}$$

where d_t denotes the bed diameter.

48

The local solids volume concentration at height h is then

$$c_v = (1 - \varepsilon_b) \cdot c_{vd} \tag{3-53}$$

with c_{vd} according to WERTHER & WEIN (1994)

$$c_{vd} = c_{vmf} \cdot \left(1 - 0.14 \cdot Re_p^{0.4} \cdot Ar^{-0.13}\right) \tag{3-54}$$

In this correlation c_{vmf} is the solids volume concentration at minimum fluidization. Re_p and Ar are defined as followed

$$Re_p = \frac{(u - u_{mf}) \cdot d_p \cdot \rho_f}{\eta_f} \tag{3-55}$$

$$Ar = g \cdot \frac{d_p^3 \cdot (\rho_s - \rho_f) \cdot \rho_f}{\eta_f^2} \tag{3-56}$$

The mass of solids in the bottom bed can be calculated from

$$m_b = A_t \cdot \rho_s \cdot \int_0^{H_b} c_v \cdot dh \tag{3-57}$$

In the upper dilute zone the solids volume concentration decreases with height. KUNII & LEVENSPIEL (1991) suggest an exponential decay where the solids volume concentration at the vessel exit is given by:

$$c_{v,e,i} = c_{v,i}^* + \left(c_{v,be,i} - c_{v,i}^*\right) \cdot e^{-a \cdot H_f} \tag{3-58}$$

The solids volume concentration $c_{v,be,i}$ at the bed surface for class i is given by the modeling of the bottom zone. H_f is the height of the freeboard between the vessel top and the height of the bottom zone which can be concluded from eqn. (3-57) for a given mass of the bottom bed. The constant a is an empirical parameter whose value is given in literature (KUNII & LEVENSPIEL, 1991). The values vary between 2 and 12. For calculations in this work 3 is often used for a.

$c_{v,i}^*$ is the solids volume concentration which would be reached with height when an infinite height is assumed. It is given by the mass flow of solids which are entrained through the interface between bottom bed and freeboard by the same method as used for the entrainment of solids from a bubbling fluidized bed. From there the solids volume concentration at the bed surface can be calculated as followed:

$$c_{v,i}^* = \frac{k_{\infty,i} \cdot \Delta Q_{3,b,i}}{\rho_{s,i} \cdot u} \tag{3-59}$$

The average solids volume concentration $\bar{c}_{v,i}$ can be yield by integration of eqn. (3-58).

$$\bar{c}_{v,i} = c_{v,i}^* + \frac{c_{v,be,i} - c_{v,e,i}}{a \cdot H_f} \tag{3-60}$$

The mass in the freeboard is calculated from

$$m_{f,i} = H_f \cdot A_t \cdot \rho_{s,i} \cdot \bar{c}_{v,i} \tag{3-61}$$

Together with eqn. (3-57) the total mass of inventory in the riser is the summation over all classes and the sum of mass in bottom bed and freeboard.

$$m_{tot} = \sum (m_{b,i} + m_{f,i}) \tag{3-62}$$

Via eqn. (3-59) the masses of bottom bed and freeboard are coupled by the dependence of $c_{v,i}^*$ on the mass fraction of class i in the bottom bed. For closing the mass balance the total mass and PSD in the bottom bed has to be calculated iteratively to meet together with the calculated mass in the freeboard the given total mass of riser inventory.

When the mass balances are solved the elutriated mass of class i out of the riser within one time step is given by the gas flow rate and the solids volume concentration at the top of the riser.

$$m_{e,i} = c_{v,e,i} \cdot \rho_{s,i} \cdot A_t \cdot u \cdot \Delta t \tag{3-63}$$

3.4. Additional sub models

3.4.1. Calculation of minimum fluidization velocity

The minimum fluidization velocity is a critical velocity where the status of a fixed bed is changed to fluidization. With the Ergun equation the flow through fixed beds can be calculated and with the assumption of minimum fluidization to be a special case of fixed bed flow the velocity under these conditions can be approximated. With the usage of the Ergun equation it is further assumed, that the particles are from spherical shape and the surface mean diameter d_p is taken as characteristic average diameter of the bulk. The voidage ε_{mf} at minimum fluidization is assumed to be 0.44 for most materials.

$$\frac{\Delta P}{H_{mf}} = 150 \cdot \eta_f \cdot \frac{(1-\varepsilon_{mf})^2}{\varepsilon_{mf}^3} \cdot \frac{1}{d_p^2} \cdot u_{mf} + 1.75 \cdot \frac{(1-\varepsilon_{mf})}{\varepsilon_{mf}^3} \cdot \frac{\rho_f}{d_p} \cdot u_{mf}^2$$

$$(3\text{-}64)$$

At these conditions the pressure drop of the fixed bed calculated by eqn. (3-64) is equal to the pressure drop of a fluidized bed given by simple force balance.

$$\frac{\Delta P}{H_{mf}} = (\rho_s - \rho_f) \cdot (1-\varepsilon_{mf}) \cdot g \qquad (3\text{-}65)$$

With eqn. (3-65) set equal to eqn. (3-64) the minimum fluidization velocity u_{mf} can be calculated.

3.4.2. Calculation of terminal velocities

In the modules for the modeling of the effects in the cyclone and for the entrainment in the fluidized bed the terminal velocities of the particles under the local flow conditions are needed. The terminal velocity for a particle of size x_i in a general flow regime under normal gravity can be described by

$$w_{s,i} = \sqrt{\frac{4}{3 \cdot C_w} \cdot \frac{\rho_s}{\rho_f} \cdot x_i \cdot \left(\frac{\rho_s - \rho_f}{\rho_s}\right) \cdot g} \qquad (3\text{-}66)$$

For the determination under defined flow conditions the value for the drag coefficient C_w has to be known. BRAUER (1971) cites an approximation of C_w for spherical particles for Reynolds numbers in the range of $0 - 200,000$ developed by KASKAS (1964).

$$C_w = \frac{24}{Re} + \frac{4}{\sqrt{Re}} + 0.4 \qquad (3\text{-}67)$$

with the definition of Re

$$Re = \frac{w_{s,i} \cdot x_i \cdot \rho_f}{\eta_f} \qquad (3\text{-}68)$$

With this set of equations the terminal velocities of particles can be determined iteratively for the whole range of conditions which are desired to be modeled. For the centrifugal flow field in the cyclone the gravity acceleration g is then replaced by the local centrifugal acceleration z.

4. Experimental

The experimental section of this work is divided into three major parts. The first part includes the generation and determination of the ash particle size distribution for the attrition assessment of ash and the modeling of a circulating fluidized bed combustor (CFBC). The second part describes the experimental methodology to determine and check the hypothesis of the single sources of attrition. These experimental set ups are based on the methods and techniques of the former works from WERTHER & XI (1993) and REPPENHAGEN & WERTHER (2000) with modified evaluation methods. The third part consists of experiments at laboratory and pilot-plant scale systems to validate the complete model. Two different systems were used. For discontinuously operated systems a bubbling fluidized bed with FCC as bed material was applied and as example for a multicomponent and continuously operated system a pilot plant scale circulated fluidized bed combustor was taken.

4.1. Generation and determination of the primary ash particle size distribution (PAPSD)

As described in the previous chapter regarding the modeling strategy for the description of the development of the particle size distribution in a combustor the burnout and combustion of fuel is not been modeled but the processes there after. Therefore the starting particle size distribution of the remained ash particles is needed as an input parameter. A lot of procedures are thinkable how to get it and it is also a question of definition. Here the so called PAPSD is defined to be the particle size distribution directly after the complete burnout and primary fragmentation of loose agglomerates.

Boëlle et al. (2002) give a good overview of the different procedures. It is very likely that the best results could be obtained when in the method of generation of the ash to measure its PAPSD the conditions are realized like in a fluidized bed combustor.

In this work two different fuels were used namely polish coal and pellets of dried sewage sludge. The determination of the PAPSD is done in two different ways depending on the fuel.

4.1.1. PAPSD of sewage sludge pellets

The sludge is burnt under moderate fluidizing velocities (1 m/s) in the pilot plant scale combustor. The combustion is done batchwise with batches of 2 kg. It is fed as a batch into the preheated empty combustion chamber and burnt at a temperature of 850°C which is controlled by adjusting the oxygen content in the fluidizing air. When the oxygen concentration at the exit of the combustor reaches the same value of the fluidizing air the combustion is considered to be complete and the material is drained through the bottom drain. The particle size distribution of the ash obtained is measured by sieve analysis and assumed to be the PAPSD.

4.1.2. PAPSD of coal

Contrary to the sludge the coal has to be treated in a different way. The sludge has an ash content of about 36 wt.-% (raw) the coal of only 7 wt.-% (raw). Additionally, the coal produces a very fine ash which would vanish very fast with the off-gas and will be lost. Therefore the coal is treated in a more complex way. The generation of a coal PAPSD is separated into two steps.

1. Combustion of the fuel in an oven without any mechanical stress (only thermal stress) to produce ash.

2. Application of stress to simulate the step of primary fragmentation of loose agglomerates.

The ash produced by oven combustion remains intact and in the original shape of the mother coal particles. In order to produce primary ash particles stress must be applied to separate loose ash agglomerates. Such a procedure applying stress through sieving for the generation of the PAPSD was suggested by BoëLLE et al. (2002). This procedure involves the sieving of ash for 30 min to reduce ash agglomerates to primary particles. Other methods of particle stressing which more closely resemble a fluidized bed can also be used, such as, ash fluidization at different temperatures and with additional bed material. Since resulting ash distributions vary and predictive models for ash inventory purposes are sensitive to this variation, it makes sense to select a method that tries to mimic the conditions of a fluidized bed combustion environment. The procedures tested are now presented and explained in detail.

First step: Ash generation by oven combustion

The following procedure is followed for temperature levels ranging from 650 °C to 950 °C. Approximately 50 g of coal was spread over the rectangular metallic oven trays. Care is taken to disperse the particles, so as to minimize groups of particles burning together. The oven is preheated to the desired temperature and the trays inserted. A close eye is kept on the samples during the period of devolatilization, approximately 5 minutes, and then the samples are left for about one hour or until all the coal is combusted. Combustion requires different lengths of time depending on the size of the particles, generally less time for smaller particles and more time for larger ones. To enhance the process an air supply is inserted through the oven door to increase air circulation. Combusting coal is visually detectable as glowing particles so that the sample is considered completely combusted when no particles were left glowing. If there is any question as to uncombusted char the sample should be monitored for an extra few minutes for visual changes. Once the particles are completely combusted the trays are removed and allowed to cool down.

Second step: Application of mechanical stress

Sieve PAPSD procedure

The ash particles from the oven combustion are sieved to break apart loose ash agglomerates. Sieving times of 10 min, 20 min and 30 min were tested for variation in the resultant ash distributions, but, no significant variation was found. In the end the sieve time was set to 30 min to be consistent with the procedure developed by BOËLLE et al. (2002). Longer sieving times should be avoided to reduce attrition of primary ash particles. The sieves should be selected depending on the size range of ash so that the analysis can be performed with one stack of sieves. Sieving a coarse fraction and then repeating the process for the remaining fines is avoided so that all fractions were stressed equally.

Fluidized bed PAPSD procedure

The stress by fluidization is applied in the fluidized bed sketched in figure 4-1. The facility is used for fluidization under ambient conditions and under combustion temperatures.

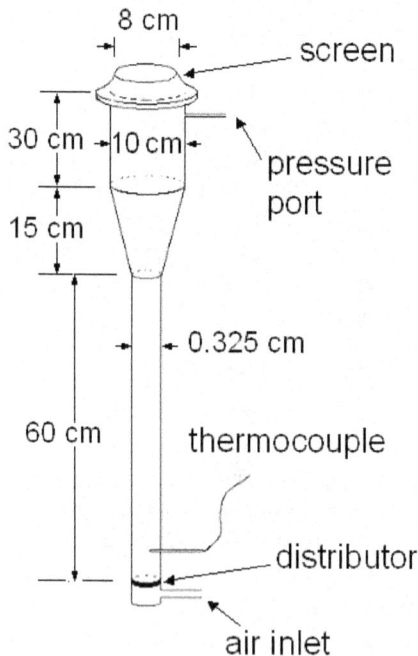

Figure 4-1: Fluidized bed for fluidizing of PAPSD-particles (the numbers are valid for inner diameter)

An amount of ash particles sufficient to provide a bed height twice that of the apparatus diameter, i.e. 200 ml or 50 g ash, is put into the facility. On top of the column a metal screen is fixed. The screen had a mesh width of 10μm and is used to avoid loosing of fine particles being entrained. The aim is to fluidize the material strongly enough to apply nearly the stress like in an industrial process; here 1 m/s should be reached. Under these conditions a lot of fines are elutriated and caught in the screen. This leads to a high pressure drop of the screen. Therefore the fluidizing velocity was increased in steps of 20 cm/s until 1 m/s was reached. The target velocity is then kept constant for 15 min. After every increase the screen is cleaned when the pressure drop gets to high. Here the assumption is made that once the particles are elutriated they have already undergone primary fragmentation and are part of the PAPSD which is to be determined. Then the ash is removed from the apparatus and weighed. The collected ash is then sieved for 10 minutes to obtain the resulting cumulative mass distribution.

In industrial processes the fuel is usually not burnt alone but in mixture with an inert bed material, ash or sand. The amounts of fuel or fresh ash, respectively, are of the order of magnitude of several percent of the bed inventory. To mimic this circumstance the procedure of fluidizing the ash is modified by mixing the ash with sand in the range of 10 to 35 wt.-% ash in the total mass. This mixture is then treated in the same way as explained above. With the knowledge of mass ratio and particle size distribution of sand which is assumed to remain the same, the particle size distribution of the ash in the mixture can be calculated from the determined particle size distribution of the mixture after fluidization.

4.2. Determination of the attrition parameters

As seen from chapter 2 there are three dominating sources of attrition in a fluidized bed system. The particles are attrited by mechanical stress in the vicinity of gas jets at the gas distributor and by vigorous movement within the bed induced by rising bubbles. The third source is the cyclones often used as separator for the particles recovery.

To investigate the attrition behavior of particles at these sources experimental set ups are needed where they can be studied independently of each other. WERTHER & XI (1993) have developed a method for the first two sources jet and bubble-induced attrition and REPPENHAGEN & WERTHER (2000) for the attrition in cyclones. These set ups will shortly be described in the following. Furthermore the modifications made for the investigation of the dependence on the stress history experienced by the particles will be outlined.

4.2.1. Attrition assessment for jet and bubble-induced attrition

The experiments regarding attrition inside of a fluidized bed are carried out in the facilities sketched in Figure 4-2. Two set ups are available for the investigation of bubble-induced attrition. The left one has an inner diameter of 0.2 m and the right one of 0.05 m. The two risers are equipped with an expansion at the top to operate as a gravitational separator by decreasing the gas velocity. The ratio of diameter of the lower part to the diameter of the separator is the same for both set ups.

At the beginning of the experiment a fresh portion of non pre-stressed solids is put into the fluidized bed. The large riser is filled with about 5 kg of material and the smaller one with about 0.25 – 0.5 kg. This leads then to a bed height at minimum fluidization of 0.15 – 0.25 m depending on solids bulk density.

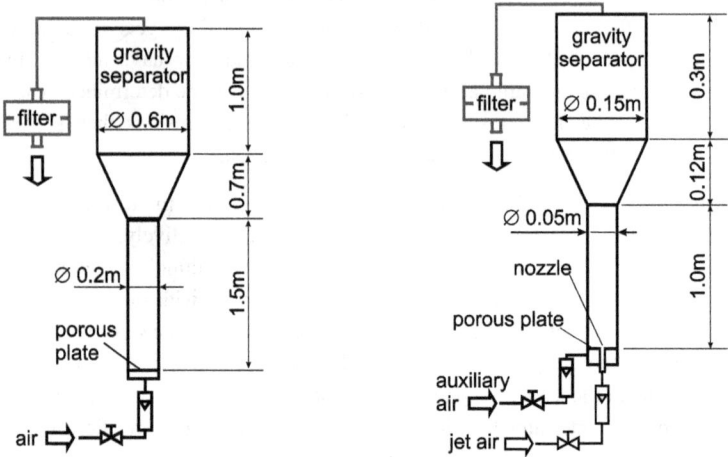

Figure 4-2: Set ups for jet and bubble-induced attrition: left set up for only bubble-induced attrition and right for both (diameters are inner dimensions)

The bed material is fluidized via a porous plate with pressurized air. The material is sieved before the experiment to contain only particles bigger than the cut size of the separator at the operating velocity.

For running an experiment to jet-induced attrition data the jet surrounding fluidized bed is fluidized via the porous plate very gently just to assure a homogenous mixing of the material. The jet is realized by a single hole in the middle of the porous plate (cf. Figure 4-3) with separate air supply.

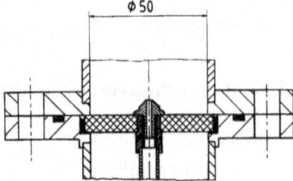

Figure 4-3: Distributor plate for jet-induced attrition assessment

The entrained particles which are abraded from the mother particles are separated from the gas by fabric filters. Both of the plants have a filter system consisting of two filters with a two-way valve between. By this configuration a continuous operation is possible without shutting down to change the filters. For the bigger plant filter bags are applied which are weighed before and after the usage and then cleaned by back flushing with pressurized air and used again. The right plant is using paper filters which were dried and weighed before and after the usage and used only once.

The filters are changed or cleaned in varying time intervals. The first change takes place after the first half hour and then the time interval will be increased up to a whole day or maximum 36 hours. The mass of the solids on the filters is assumed to be only particles which are originating from the surface of the mother particles by abrasion. The mass is then divided by the width of the time interval and an average mass flow of produced fines is obtained and related to the middle of the time interval.

4.2.2. Attrition in cyclones

The procedure for the attrition in cyclones is basically the same as for the jet or bubble-induced attrition. A portion of solids is stressed by passing through a cyclone several times and the fines produced are caught on a filter and weighed. The set up for the assessment is shown in figure 4-4.

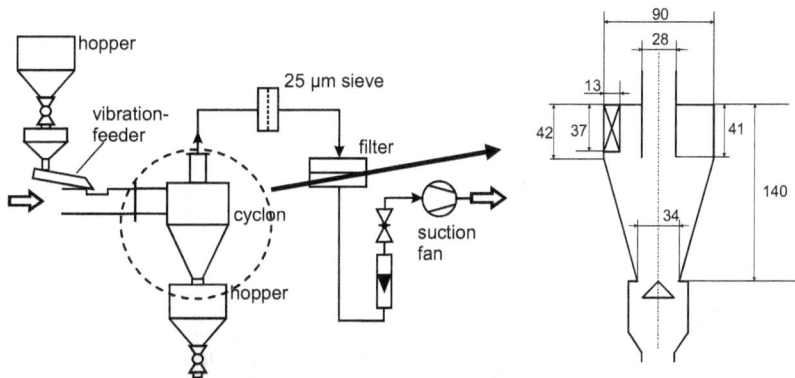

Figure 4-4: Flow sheet and cyclone dimensions of experimental set up for assessment of attrition in cyclones (dimensions are given in mm)

For one pass through the cyclone the solids are fed into the inlet pipe of the cyclone via a vibration feeder. By running through the cyclone they are undergoing attrition and the mother particles are separated and collected in the hopper below the underflow of the cyclone. The fines produced which are usually smaller than 5 µm are entrained from the cyclone with the off-gas through the overflow and then caught on the filter. Before the filter misled coarse particles are separated from the fines by a 25 μm sieve. After one pass through the cyclone the solids are taken from the collecting hopper and put back into the feeding hopper and the particles are stressed for the next time.

The filters are the same paper filters than for the jet-induced attrition tests and are used for up to 4 runs. After the runs the filter is taken out dried and weighed. The mass increase of the filter is assumed to be produced fines and divided by the number of corresponding runs and the sum of the masses fed during these runs. A mass related attrition rate with the dimension of kg/kg/run is yield. This value is then the average attrition rate of these runs.

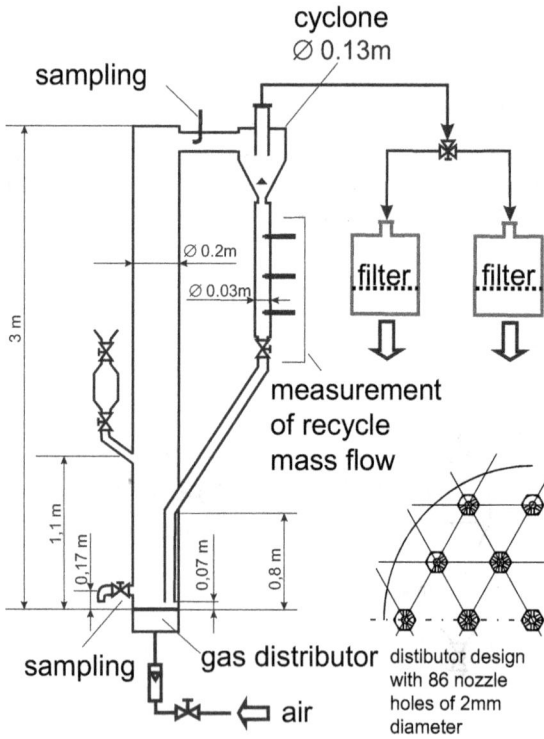

Figure 4-5: Bubbling fluidized bed test plant

4.3. Validation of model

After the determination of the model parameters such as input PSD and attrition parameters, experiments have been conducted at two different fluidized bed systems. These experiments yield data to be compared with model calculations. The model provides information about mass flow and PSD at different locations in the system. In the case of a circulating fluidized bed the axial profile of the solids volume concentration is calculated. These characteristic values are measured in the experimental set ups to be compared with the model calculations.

4.3.1. Set-up of the bubbling fluidized bed at batch-wise operation

A batch-wise operating FCC fluidized bed is chosen to investigate the development of attrition in the start-up period of such a system. The system is operated under bubbling fluidized bed conditions and has the configuration presented in figure 4-5.

The fluidized bed is equipped with a distributor built with 86 nozzles. The nozzles have diameters of the outlet openings of 0.002 m. The recirculation cyclone of the system is in contradiction to industrial bubbling fluidized beds not inside the riser and an expansion joint is not installed. With this configuration it is easier to measure the particle flow and size distribution of the entrained flow from the bed. The recirculation pipe is conducted back into the riser above the bed surface and then diving through the bed surface down into the bed and ending 0.07 m above the nozzles. In the inlet channel of the cyclone a port for the suction probe is provided where solid samples can be taken at different vertical positions in the channel.

In the downcomer below the cyclone some capacitance detectors are installed at different heights for the determination of solids recirculation rate. When the ball valve is closed the solids are accumulating in the down-comer and with the help of the detectors the time can be measured which will be needed to fill the downcomer to the height of the detector. The down-comer was calibrated with the solids used before.

For taking samples from the bed a bottom drain is provided where a bigger portion of particles is taken and split by a sample splitter down to a small sample of a few grams. The rest is given back to the system by a solids lock at the height of 1.1 m.

The solids lost from the system which could not be kept in the system by the cyclone are leaving with the overflow and caught in a fabric bag filter. The filter system is also configured with two filters which can be used alternately. By weighing the filters and dividing the increase in mass by the time interval the average loss rate is obtained and referred to the middle of the time interval.

Solid sampling for determination of local mass flow and PSD in the cyclone inlet

In the system presented above the solid sampling system described here has been applied to determine mass flow and PSD of solids in the inlet duct to the cyclone.

For solid sampling from gas-solid flows suction probes are used. Figure 4-6 shows the flow sheet of the system. By a pipe introduced into the flow a part of the flow is sucked and led through a small cyclone. The cyclone is separating the particles from the gas. The overflow of the cyclone is connected via a filter with the inlet of the pump. After the pump the gas flow is divided by two flow meters with needle valves into an exhaust stream and a recirculation stream. The recirculating gas is led to the tip of the probe and mixed with the sample stream. This leads to an acceleration of the gas flow from the probe to the cyclone which avoids avoid any settling of particles in the pipes. The ratio of exhaust stream to recirculation is adjusted, that the incoming gas velocity at the inlet of the probe is nearly the same than the surrounding gas velocity in the sampling region.

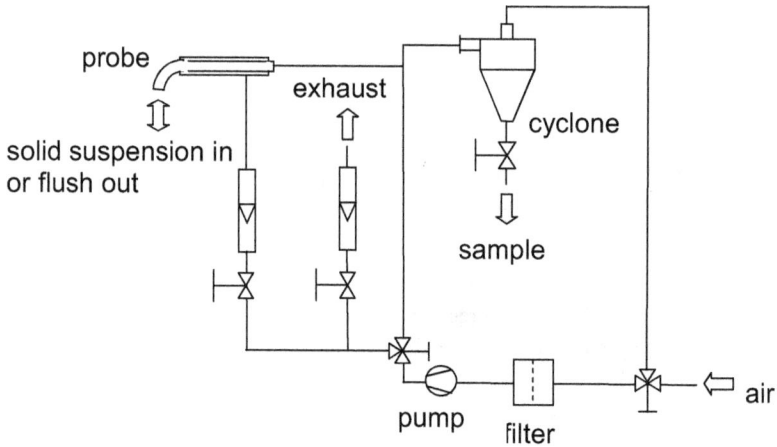

Figure 4-6: Flow sheet for solids sampling in gas-solid flows.

For the second operation mode air is drawn in from the environment and led through both outside ports of the probe to the tip in order to flush the probe. In this mode the probe can be cleaned in place.

By monitoring the sampling time and weighing the collected solids samples a mass flow rate can be calculated and divided by the inlet area of the probe converted to an area specific mass flow rate.

4.3.2. Set-up of the circulating fluidized bed combustor (CFBC)

The second system which is investigated and validated is the pilot-scale circulating fluidized bed combustor given in Figure 4-7. The CFBC riser has an inner diameter of 0.1 m and a height of 15 m. It is equipped with electrical heating jackets and an air preheater to level out heat losses. After the cyclone gas is cooled down to about 473 K and the fly ash is separated by fibric filters. Fly ash is also collected in the flue gas cooler and samples can be taken. At the exit from the cyclone the gas concentrations are monitored for the control of the unit.

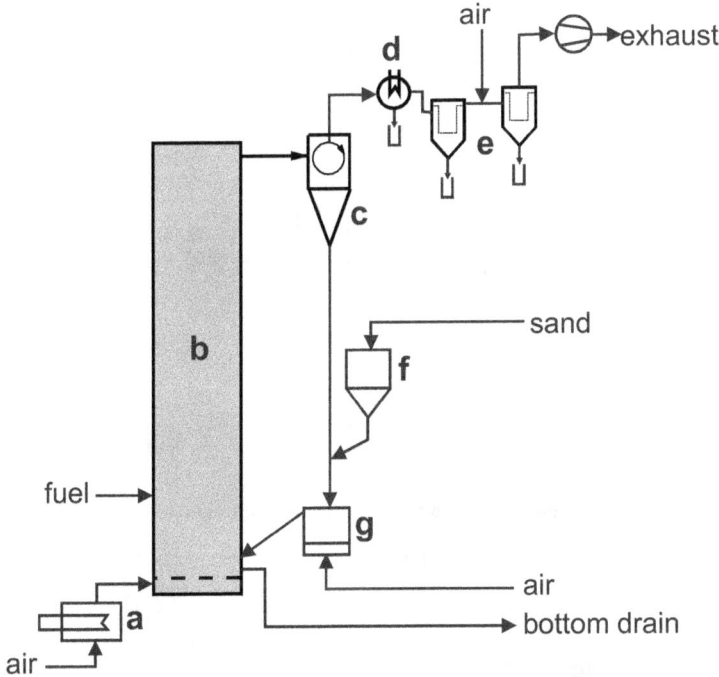

Figure 4-7: Sketch of pilot plant; a - air preheater; b - combustion chamber; c - cyclone; d - cooler; e - filter; f - sand lock; g - siphon

The combustor is operated at constant temperature (here 1125 K) and a constant bed inventory which means a constant pressure drop of the riser. In the experiments of this work the pressure drop was kept constant at 7500 Pa with an accuracy of 7%. To keep the mass of inventory constant a sand lock is provided in the return line where sand as bed material can be added when the pressure drop is too low. At high pressure drop bed material can be withdrawn through the bottom drain.

<u>Fuel feeding system</u>

The fuel is fed via a screw feeder to a short down-comer the lower end of which is attached to the riser at a height of 4.2 m above the distributor. Experiments were conducted with two different fuels, namely polish coal and pellets of dried sewage sludge. Each fuel needs regarding its feeding characteristic an own screw feeder and for some experiments the fuels were used in mixture for co-combustion. Therefore the screw feeders were attached to the down comer via a Y-shaped connection as shown in Figure 4-8.

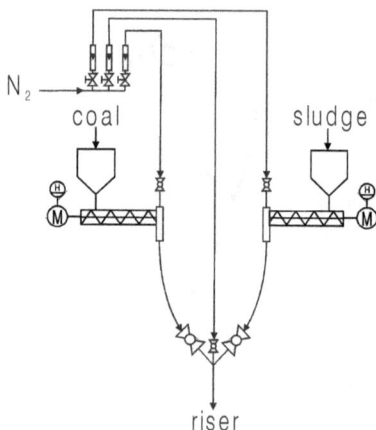

Figure 4-8: Feeding system for coal and sludge

This configuration allows single combustion with fast step changes from one fuel to another one and for combustion of arbitrary mixtures of the two fuels. Different fuels have different ashes which also influences particle size distributions of particles in the system.

<u>Solids volume concentration and temperature profile measurement</u>

Along the riser pressure taps and thermocouples are distributed. With the axial pressure profile the local apparent solids volume concentrations can be calculated. WERDERMANN (1993) calculates the apparent solids volume concentration from the pressure difference, Δp_M over a height segment as follows

$$c_{v,app.} = \frac{\Delta p_M}{g \cdot \Delta h \cdot (\rho_s - \rho_f)} + \frac{\rho_{f,L} - \rho_f}{\rho_s - \rho_f} \qquad (4\text{-}1)$$

Where ρ_s is the density of the solids, ρ_f the density of the gas inside the riser at operating temperature and $\rho_{f,L}$ the gas density at ambient conditions or at the temperature outside the combustion chamber, respectively.

Additional probe ports are located at heights of 6.2, 10.5 and 14.5 m above the distributor. Through these ports solid samples from the riser center could be taken via the solid sampling system described in following section or a heat transfer probe could be installed to detect changes in bed to wall heat transfer.

Detection of changes in bed to wall heat transfer

When the PSD of the bed inventory is changed this has a strong influence on the distribution of the particles within the riser. A coarse bed material leads to low solids volume concentrations in the upper dilute region and vice versa. The bed-to-wall heat transfer is direct related to the local solids volume concentration. A decrease of the cross-sectional average solids volume concentration at one height causes a decrease of heat transfer coefficient at this height (BREITHOLZ et al., 2001). Therefore a heat transfer probe has been developed and was used in the experiments for sensing changes in the bed-to-wall heat transfer.

The set up of the probe is shown in figure 4-9. In this probe the temperature change of a constant water flow is measured by a high precision temperature measurement system (QuaT by Heraeus, Hanau) with a precision of 0.1 K. A change of heat transfer will lead to a different heating of the water at constant conditions. During the experiments the probe was located at heights of 6.2, 10.5 and 14.5 m, respectively, above the distributor.

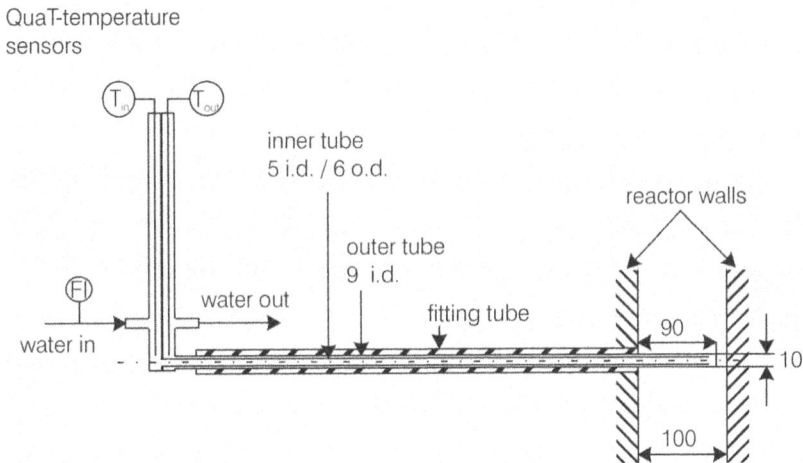

Figure 4-9: Heat transfer probe

From the results of the heat transfer probe measurements heat transfer coefficients can be approximately determined by the following relationship:

$$\dot{Q} = \dot{m}_{water} \cdot c_{p,water} \cdot (T_{out} - T_{in}) = k \cdot A \cdot (T_{riser} - \overline{T})$$ (4-2)

where

$$\overline{T} = 0.5 \cdot (T_{out} + T_{in})$$ (4-3)

Heat transfer coefficient k can then be approximated by:

$$k_b = \frac{\dot{m}_{water} \cdot c_{p,water}}{A} \cdot \frac{T_{out} - T_{in}}{T_{riser} - \overline{\overline{T}}}$$ (4-4)

It has to be outlined that this system is not suitable for the determination of precise heat transfer coefficients. The coefficients obtained with these measurements are averaged ones for a heat transfer pipe across the riser cross section. Since in most industrial plants the heat is usually transferred via vertical membrane walls or wing walls they have values for k that differ from those which are valid for horizontal pipes. But changes in solids distribution in the riser by changes in the heat transfer at the probe and the order of magnitude of changes in heat transfer can be estimated.

5. Results

In this chapter the evaluation of the experiments described in chapter 4.2 is presented. These experiments deal with the determination of model parameters which will be needed for the application of the model. With these parameters the operation of the two presented fluidized bed systems (cf. 0) will be simulated in the following chapter and the calculation results will be compared with experimental findings. The intention is to determine the model parameters independently from the system which is intended to be described.

5.1. Time-dependence of attrition

In figure 5-1 the measured attrition rate for the bubble-induced attrition of limestone is presented as a function of time. The time-dependence of the production of fines is evident in this experiment. At the beginning a very high mass flow of fines occurs which decreases with time to a nearly constant one. The attrition rate at the beginning is about two orders of magnitude higher than at nearly steady-state. As mentioned in chapter 3.2.2 this is caused by the surface properties of fresh particles. The fines produced at the beginning originate from breaking edges and cracks. After a while these rough edges have vanished and the surface is smoothened to a natural roughness which will remain constant. In this condition the attrition rate of the particle will not change with time anymore. Just the production of fines will decrease with decrease of particle size (cf. chapter 3.2.1).

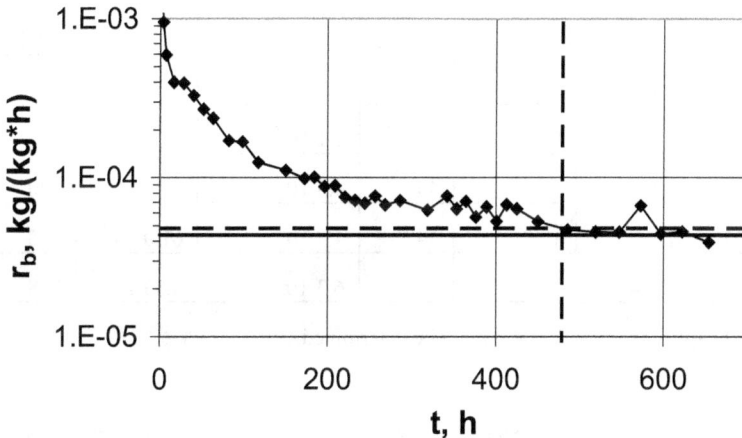

Figure 5-1: Time-dependence of bubble-induced attrition and determination of t_b* (limestone at u = 0.5 m/s).

For the model presented in chapters 3.2.1 and 3.2.2 some parameters have to be extracted from these results. The first one which is needed is the attrition rate r_∞ for complete smoothened and rounded particles. When nearly no change in attrition rate is observed the average of the last $4 - 7$ measured values are taken as r_∞ (marked as the horizontal solid line). This evaluation is following the method suggested by WERTHER & XI (1993).

In order to describe the curve mathematically a fixed point is needed. Two points are imaginable which are the attrition rate at $t = 0$ or the time when the particles a just fully smoothened and reached the attrition rate r_∞. Both points can not be estimated with acceptable accuracy from the plot. Instead a time $t_b{}^*$ is defined to be the time when the attrition rate is 10% higher than the steady-state attrition rate r_∞. This is visualized by the dashed lines in figure 5-1.

With r_∞ being equal to 4.3 10^{-5} kg/kg/h in this example $t_b{}^*$ is determined as 480 h to reach the attrition rate of $1.1 \cdot r_\infty$. This procedure is applied to all bubble-induced attrition tests where fresh material is used at the beginning. Table 5-1 contains a summary of the characteristic values obtained in the experiments bubble-induced attrition. Besides the operating conditions, defined by fluidizing velocity, the surface mean diameter and the minimum fluidizing velocity of bed material is given in addition to the steady-state attrition rate $r_{b,\infty}$, the characteristic time $t_b{}^*$ and the attrition coefficient C_b.

Table 5-1: Characteristic results from bubble-induced attrition experiments.

material	u, m/s	u_{mf}, cm/s	d_p, µm	$r_{b,\infty}$, kg/kg/h	C_b, s^2/m^4	$t_b{}^*$, h
limestone	0.6	0.9	19	7.2 10^{-5}	6.0 10^{-3}	*
	0.4	0.9	15	2.1 10^{-5}		*
	0.5	5.3	49	4.3 10^{-5}		480
Sludge ash	0.5	12.2	44	4.8 10^{-4}	61.0 10^{-3}	105
FCC-catalyst	0.4	1.4	57	6.3 10^{-6}	0.53 10^{-3}	358

* tests with spent material. Value cannot be determined.

The development with time of the attrition rate for the jet-induced attrition looks similar to the one for bubble-induced attrition (cf. figure 5-2). Contrary to the bubble-induced attrition rate the jet-induced attrition rate cannot be measured independently. The fact that the bed surrounding the jet has to be fluidized causes additional attrition by bubble movement (WERTHER & XI, 1993). For the attrition rate at nearly steady-state conditions it is the sum of jet-induced attrition rate and the bubble-induced one. Therefore the jet-induced attrition rate for rounded particles $r_{j,\infty}$ is yield by simply subtracting the bubble-induced attrition rate $r_{b,\infty}$ from the measured total attrition rate. The attrition rate $r_{b,\infty}$ is given by eqn. (3-7) with the determined attrition coefficient C_b from previous section.

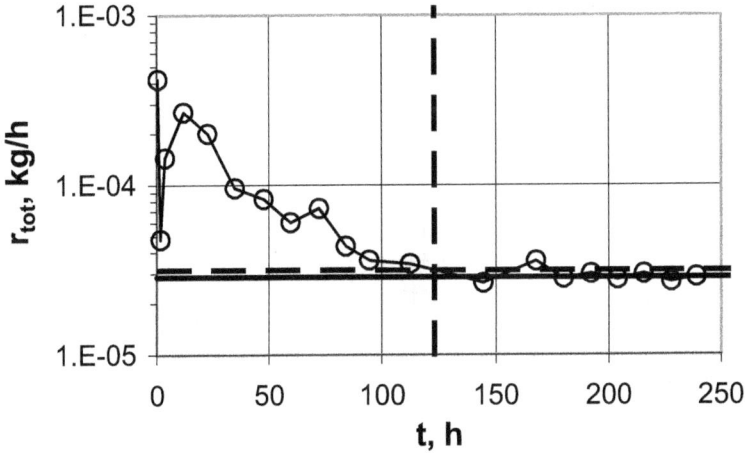

Figure 5-2: Time-dependence of jet-induced attrition and determination of t_{tot}* (limestone: u_{or} = 50 m/s; m_b = 0.25 kg).

From the time development of the attrition rate t_{tot}* is determined in analogy to the determination of t_b* for the bubble-induced attrition. It is called t_{tot}* because this value also includes the increase of stress history by bubbles in addition to the one induced by the jet. When the particles are stressed by the bubble movement their stress history changes also. These particles then enter the jet with a changed stress history than others. This results in a combined change of stress history by either bubbles or jet which can not be measured separately. With the assumption, that the stress histories can simply be added the increase of ϑ within one time step is the sum of the increases by bubbles and jet, respectively.

$$\Delta \vartheta_{tot} = \Delta \vartheta_b + \Delta \vartheta_j \qquad (5\text{-}1)$$

In detail this relationship may be expressed by

$$\Delta \vartheta_{tot} = \frac{\Delta t}{t_{tot}^*} = \frac{\Delta t}{t_b^*} + \frac{\Delta t}{t_j^*} \qquad (5\text{-}2)$$

The time step Δt is the same for all $\Delta \vartheta$ and can therefore be extracted to yield $t_j{}^*$ from the experiment.

$$t_j^* = \left(\frac{1}{t_{tot}^*} - \frac{1}{t_b^*} \right)^{-1}$$

(5-3)

In the example presented above $t_{tot}{}^*$ is determined to be 123h which results in $t_j{}^* = 165$ h with $t_b{}^* = 480$ h from previous experiment regarding bubble-induced attrition. The results and summary of operating conditions are given in table 5-2. With a value for $t_j{}^*$ being smaller than $t_b{}^*$ the acceleration of smoothening of particles is given.

Table 5-2: Results overview for jet-induced attrition

material	u_{or}, m/s	m_b, kg	u, m/s	u_{mf}, cm/s	d_p, μm	$r_{j,\infty}$ kg/h	C_j, s^2/m^3	$t_j{}^*$, h
limestone	50	0.25	0.2	1.18	26	26.1 10^{-6}	6.82 10^{-4}	165
FCC-catalyst	50	0.2	0.5	1.4	66	1.75 10^{-6}	0.122 10^{-4}	184

To complete the picture experiments conducted in cyclones were evaluated in a similar way. But instead of time the attrition rate r_c is a function of passes n_p through the cyclone. Figure 5-3 shows the decrease of attrition rate r_c with increasing number of passes n_p and visualizes again the method to estimate $n_p{}^*$. By interpolation between two measured points a real number will be obtained. Considering the fact that number of passes is always an integer the calculated value is rounded.

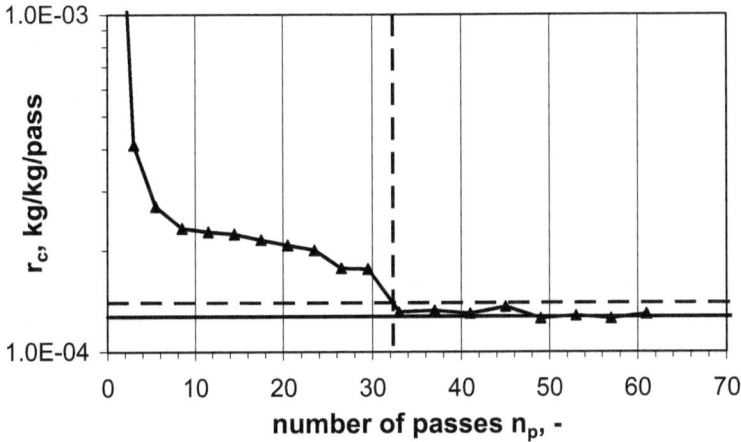

Figure 5-3: Dependence of attrition in cyclones on number of passes and determination of n_p*. (limestone: u_{in} = 18 m/s; μ_{in} = 0.3 kg/kg)

Table 5-3 gives an overview of the cyclone experiments. The number of passes needed to round the particles to a steady-state condition is in the range of 14 – 32 passes. This illustrates the strong influence a cyclone will have in a circulating fluidized bed system. In such systems the bed material might be recycled several times an hour. This could result in a domination of the cyclone on the time-dependence of the overall production of fines.

Table 5-3: Results from attrition experiments in cyclones.

material	u_{in}, m/s	μ_{in}, kg/kg	d_p, µm	$r_{c,\infty}$, kg/kg	C_c, s²/m³	n_p*, -
limestone	18	0.3	19	1.27 10^{-4}	13.5 10^{-3}	32
	14	0.3	21	1.15 10^{-4}		14
	18	0.2	25	2.11 10^{-4}		20
Sludge ash	14	0.2	39	6.15 10^{-4}	34.0 10^{-3}	17
FCC-catalyst	18	0.3	57	6.52 10^{-5}	1.9 10^{-3}	25

In chapter 3.2.2 a new parameter, i.e. the so-called stress history parameter has been defined by eqns. (3-16) - (3-18) and introduced for the description of the actual attrition rate in eqn. (3-19). The assumption is made that the relationship between the attrition rate at steady-state $r_{i,\infty}$ and the actual attrition rate r_i is given by eqn. (3-20). With eqns. (3-16) - (3-18) for each measured data point the corresponding value of ϑ can now be calculated with the estimated values of $t_b{}^*$, $t_j{}^*$ and $n_p{}^*$. In order to obtain the value for $f(\vartheta)$ which belongs to the calculated value of ϑ eqn. (3-19) is solved for $f(\vartheta)$ by dividing by $r_{i,\infty}$

$$f(\vartheta) = \frac{r_i(\vartheta)}{r_{i,\infty}} \qquad (5-4)$$

In figure 5-4 the dimensionless attrition rate according to eqn. (5-4) is plotted versus ϑ for limestone. The dimensionless attrition rates for all experiments meet one curve with a fairly low deviation. In a double logarithmic diagram the curve is approximated by a straight line for $\vartheta < 1$ and constant for ϑ running towards infinity. This is in agreement with eqn. (3-20). The slope of the straight line is then the stress history exponent b. For limestone b is found to be -0.7 which is the exponent of the included curve in the diagram; it is the black line.

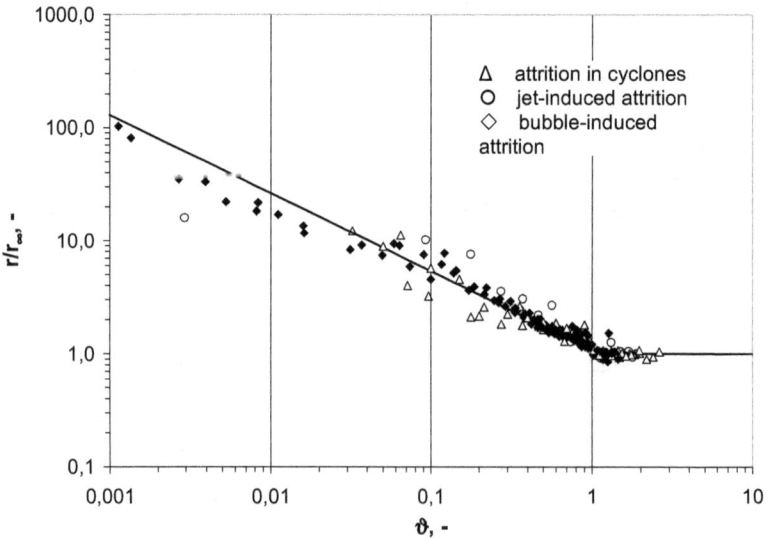

Figure 5-4: Dimensionless attrition rate of limestone as a function of stress history (the inclined line is eqn. (3-20) with b = -0.7)

70

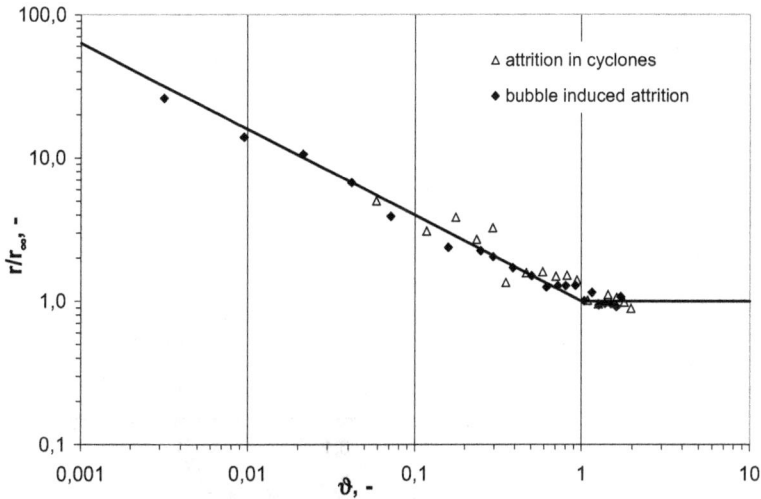

Figure 5-5: Dimensionless attrition rate of sludge ash as a function of stress history (the inclined line is eqn. (3-20) with $b = -0.7$)

In figures 5-5 and 5-6 the same relationship is found for sewage sludge ash and for FCC-catalyst particles. Eqn. (3-20) can therefore be used to describe the shape of the stress history dependence of attrition in general. It is only the stress history exponent b which has to be determined separately for the material. Probably it is as a coincidence that the exponent b for the sludge ash is with -0.7 the same than for the limestone. For the FCC-catalyst b is found to be -1.16. A high exponent means that the smoothening of particles' surface is much faster at the beginning and decreases with ϑ. Physical reasons for this behavior may be founded in the inner structure of particles as porosity, roughness of surface, inner stress or tension, etc, which were however not studied further in the present work.

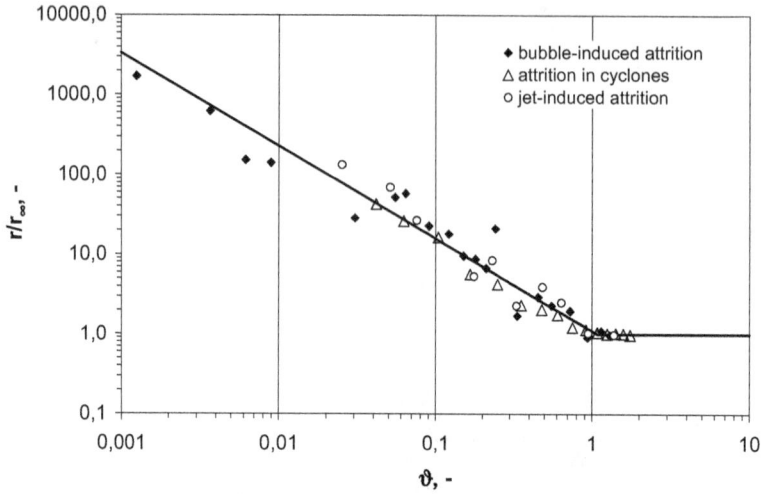

Figure 5-6: Dimensionless attrition rate of FCC-catalyst as function of stress history (the inclined line is eqn. (3-20) with $b = -1.16$)

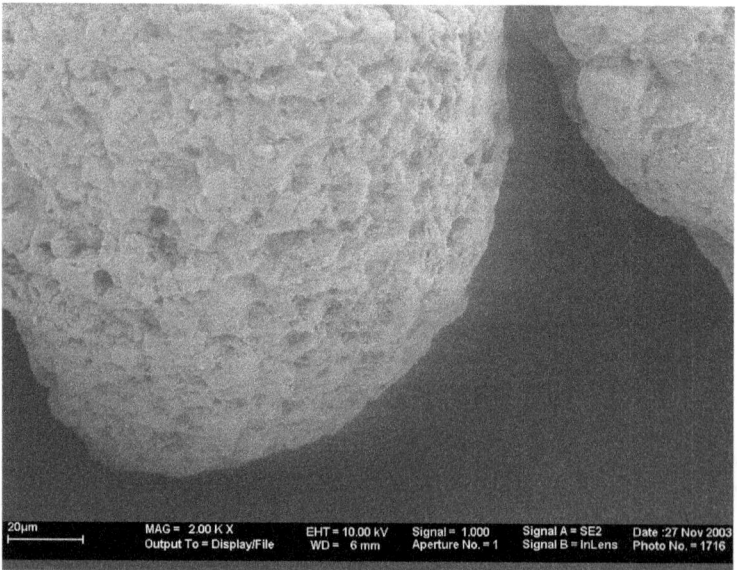

Figure 5-7: Structure and shape of limestone particles before exposed to stress

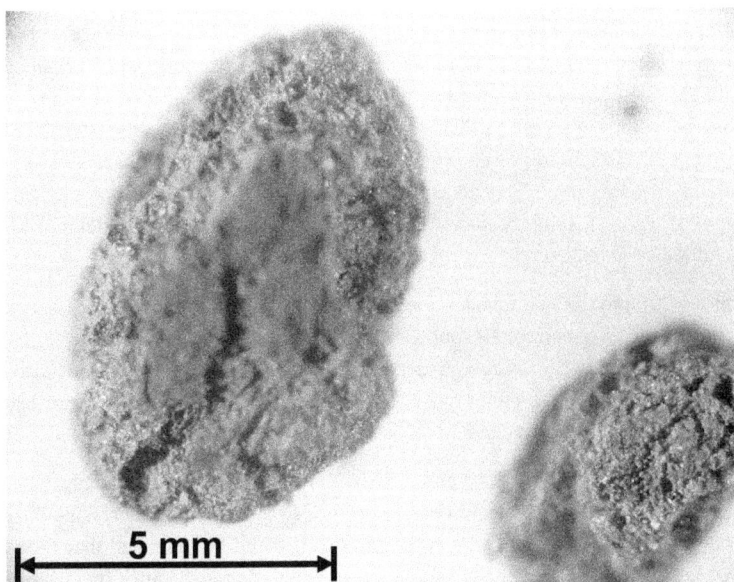

Figure 5-8: Structure and shape of ash particles from dried sewage sludge before exposed to stress

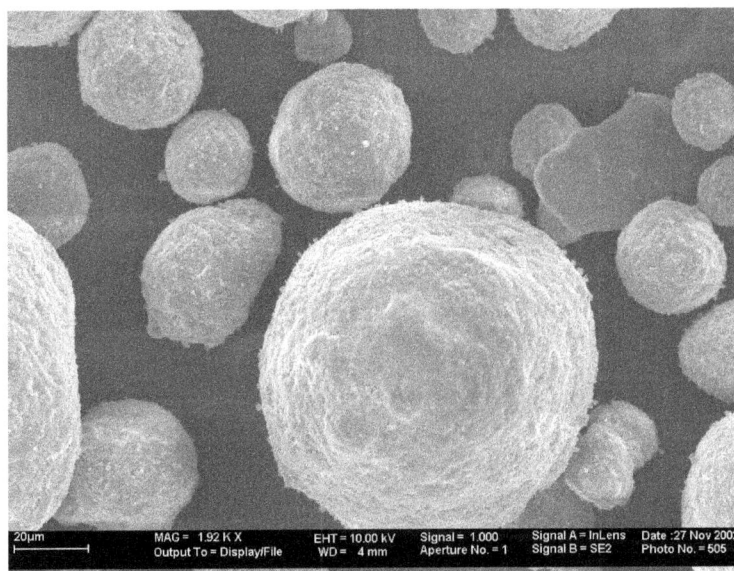

Figure 5-9: Structure and shape of FCC-catalyst particles before exposed to stress

Pictures of particles are given in figures 5-7 to 5-9. For limestone and FCC-catalyst the pictures were taken with the aid of an electron scanning microscope wherever the picture of the sludge ash is made with a usual light microscope. The surface of limestone and sludge ash is rougher than of the FCC-catalyst which could lead to the faster decrease of attrition rate at the beginning of the FCC. The rough edges of the FCC are vanishing very fast and then it takes quite long to end up with a fully smoothened particle. From the structure with coarser primary particles in the limestone and ash particles the higher attrition rates can be explained.

With eqn. (3-20) and the numerical value of the exponent b the stress history dependence of attrition can be described for all sources. But still the characteristic values t^* or n_p^* have to be determined for every source. A dependence of the values on operating conditions could not be found when the tables 5-1 and 5-3 were compared. The values are scattering around an average value for every source.

When the averaged values of t_b^*, t_j^* and n_p^* are plotted versus the attrition coefficients C_b, C_j and C_c, respectively, a significant decrease of t_b^*, t_j^* and n_p^* with increasing C_b, C_j and C_c can be observed (c.f. figure 5-10 a), b) and c)). It could be imagined that C and t^* or n_p^* are depending on the same material properties which define the attrition resistance and behavior of the particles which will result in the same qualitative behavior. For example, with a decreasing attrition coefficient C the particles have a higher resistance to attrition and it takes longer to smoothen their surface so that t^* and n_p^* are increasing.

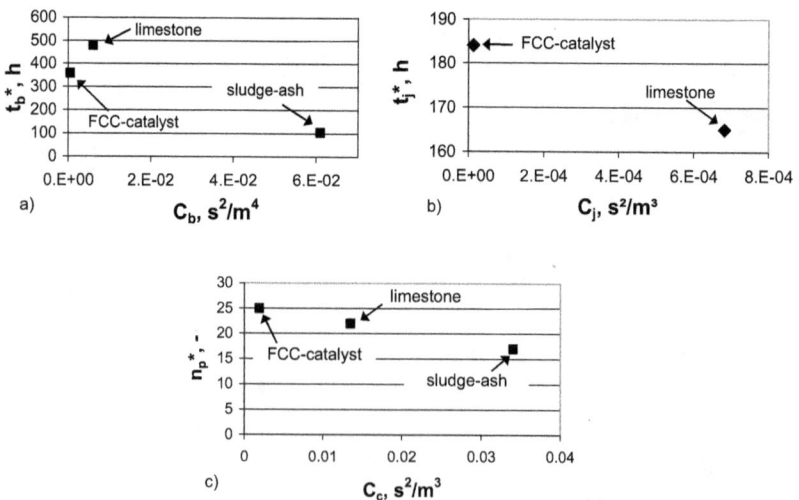

Figure 5-10: a) dependence of t_b^* on attrition coefficient C_b; b) dependence of t_j^* on attrition coefficient C_j; c) dependence of n_p^* on attrition coefficient C_c.

5.2. Attrition of mixtures

In the modeling of multi-component systems the rough assumption is made that particles of different materials with different attrition properties do not influence each other with regard to their attrition behavior. In order to examine and prove the validity of this assumption experiments were conducted with mixtures of limestone and sludge ash.

It is assumed that the contribution to the total attrition rate of a mixture of one component is given by its mass fraction on the total mass.

$$r_{tot} = \sum_k \frac{m_k}{m_{tot}} \cdot r_k \qquad (5\text{-}5)$$

The ratio m_k / m_{tot} is the mass fraction of the component in the total mass m_{tot} and r_k is the attrition rate of the component given by eqns. in chapters 3.2.1 and 3.2.2.

For a mixture of limestone and sludge ash eqn. (5-5) reads

$$r_{tot} = \frac{m_{limestone}}{m_{tot}} \cdot r_{limestone} + \frac{m_{ash}}{m_{tot}} \cdot r_{ash} \qquad (5\text{-}6)$$

Mixtures of 20 wt.-% of sludge ash with limestone were prepared and their bubble-induced attrition rate and attrition rate in cyclones measured. The sludge ash had a surface mean diameters of 44μm, whereas the limestone had one of 49μm, which is quite similar to the one of the sludge ash. Here the attrition rates at nearly steady-state conditions are determined and the validity of the above mentioned assumption examined.

Fluidization of the mixture at 0.5 m/s yields an overall bubble-induced attrition rate $r_{b,tot,\infty}$ of 6.36 10^{-8} kg/kg/s. The theoretical value according eqn. (5-5) is calculated to be 4.01 10^{-8} kg/kg/s which is within a range of a good accuracy.

For attrition in the cyclone the theoretical value of $r_{c,tot,\infty}$ = 3.94 10^{-4} kg/kg is with a factor of 2.4 greater than from the measured one of 1.67 10^{-4} kg/kg. The cyclone attrition rate of pure sludge ash is under these conditions 1.5 10^{-3} kg/kg which is one order of magnitude higher than the one of pure limestone with 1.18 10^{-4} kg/kg. Therefore it is assumed to be an acceptable agreement that the theoretical value is nearer in the dimension of the attrition rate of the limestone which contributes with 80 wt.-% most to the mass.

5.3. The primary ash particle size distribution (PAPSD)

As has been explained in chapter 3 in the modeling of a circulating fluidized bed combustor (CFBC) the reaction of fuel particles will not be described. Instead of the PSD of the fuel particles the PSD of fresh ash will be used as an input parameter. The PSD of the fuel particles directly after complete burnout and primary fragmentation is according to CAMMAROTA et al. (2002) defined as the primary ash particle size distribution (PAPSD). The PAPSD is strongly dependent on the fuel. It is influenced by fuel composition, inner structure of fuel particles and fuel PSD and has to be determined experimentally.

5.3.1. PAPSD of dried sewage sludge pellets

Due to the high ash content of the dried sewage sludge of 36 wt.-% (raw) the sludge delivers a high amount of ash per batch and the ash can be used for determination of attrition parameters (cf. previous section). The sludge pellets have a very narrow PSD with average particle size of 10 mm. These Pellets remain nearly in shape after combustion. The combustion leads to shrinkage and only some particles break apart. The sludge was burnt batch wise in a fluidized bed combustor presented in chapter 4 (cf. 4.1). Batches of some kilograms were burnt and their PSD determined by sieve analysis. In figure 5-11 the PSD's of the generated batches are plotted. Some scattering is visible due to measuring and experimental accuracy. The PAPSD for modeling in following chapter is extracted by averaging the PSDs which is given by the solid line.

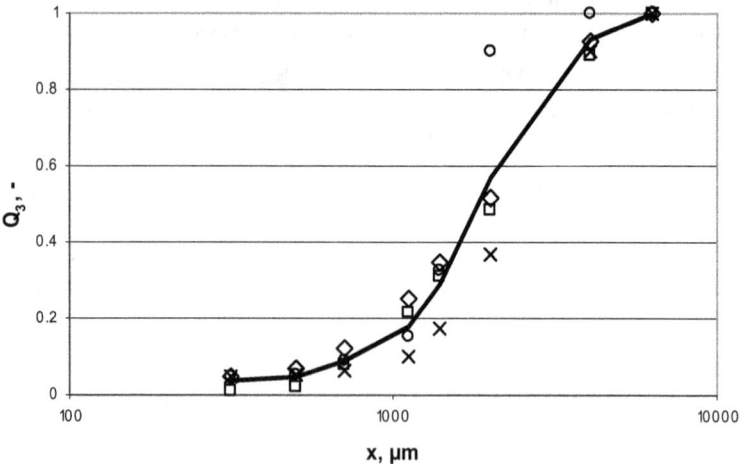

Figure 5-11: PAPSD of sewage sludge ash

5.3.2. PAPSD of coal

The polish coal used in this work has an ash content of 7 wt.% (raw) which yields ash in very low amounts after combustion. Additionally, the fuel has a very broad PSD with a lot of fines which will disappear directly during combustion or are easily elutriated when burnt under fluidizing environment. In order to obtain the complete picture of the PAPSD of the coal its generation is divided into two steps where the loss of fines is minimized. As described in chapter 4.1 the fuel is first burnt in an oven without any movement and then mechanical stress is applied by sieving or fluidizing under different conditions.

Oven combustion

During combustion in the oven the particles might already break apart due to percolation by disappearing of carbon from the particle matrix and by thermal stress or generation of inner pressure during devolatilization. The fuel particles were distributed on a metal tray in such a way that they do not touch each other so they could not melt together and hot spots do not occur. Even the oxygen supply should be the same for all. Pictures were made from the tray before and after combustion. During handling of the tray it was paid attention to not shaking it. Figure 5-12 shows one example of particles before and after combustion in the oven. The shape of the particles is nearly conserved and breaking due to devolatilization, percolation and thermal stress is not significant. Even the size of these primary ash particles remains nearly the same than of the fuel particles. But it has to be remarked that these particles can be suspected to be loose agglomerates of ash and it was observed that they break easily apart when moved. The breakage of these loose agglomerates is referred to the primary fragmentation of ash particles. The primary fragmentation is then effected by applying mechanical stress to end up with the PAPSD.

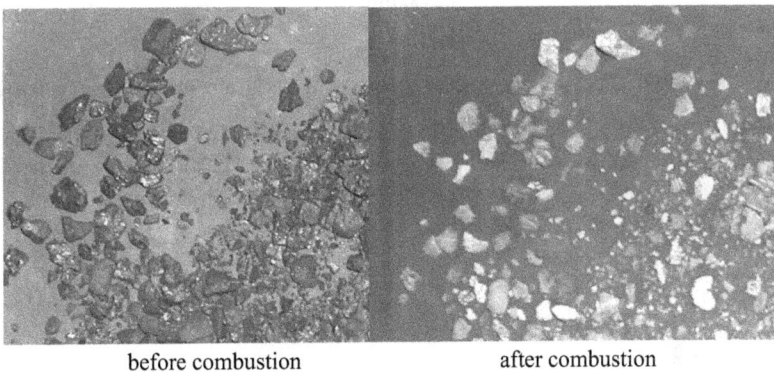

before combustion   after combustion

Figure 5-12: Fuel particles before and after combustion in muffle oven at 950°C (polish coal)

Application of stress

As described in chapter 4.1 different methods of applying stress to the ash particles are investigated. For the first investigations coal particles of size range from 2 – 5 mm were burnt end exposed to stress. Figure 5-13 leads to the conclusion that applying stress by sieving, fluidizing for 15 minutes at 25°C or 850°C yield more or less the same PAPSD which is much finer than the original fuel particles. Fluidizing the particles produces a slight finer PAPSD than sieving, but a significant difference between the PAPSD from fluidizing at 25°C and 850°C is not detectable. After the application of stress the particles were sieved anyway which is presented in figure 5-13. The fine fraction of the sieve analysis below 450μm is additionally analyzed in a laser diffraction device.

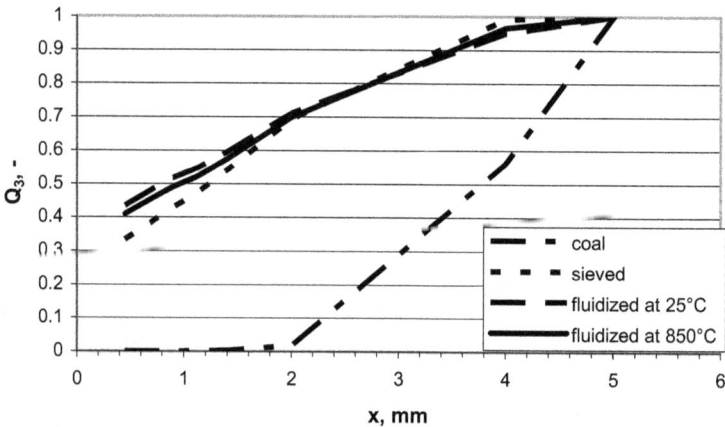

Figure 5-13: PAPSD of a coarse coal fraction generated by sieving, fluidizing at 25°C and 850°C.

Figure 5-14 shows that a significant difference between cold and hot fluidization can not be observed for the fine fraction from the sieve analysis. But in contradiction to the sieve analysis results in figure 5-13 the PSD of the PAPSD produced by sieving is slightly finer than the PAPSD from the fluidization.

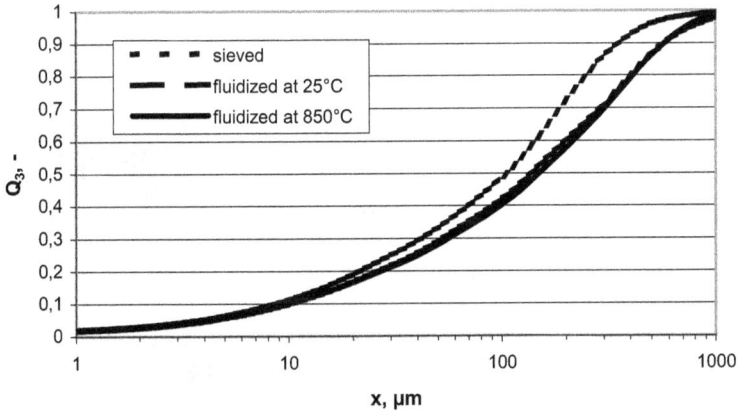

Figure 5-14: PSD of fine fraction of PAPSD originating from coarse coal fraction.

In industrial processes the ash is usually generated in the presence of inert mineral material which might have a strong influence on the size of the PAPSD particles. Therefore the application of stress under fluidization is conducted with mixtures of ash particles and quartz sand in different ratios. After fluidizing the mixture the PSD of the mixture is determined and with knowledge of the sands PSD the PAPSD is calculated and presented in figure 5-15. A decrease of ash content in the mixture leads to an increase in the fines of the PAPSD. The fraction of fines smaller than 450 µm increases from 45 % to 55 %. A significant influence of temperature could again not be observed (cf. figure 5-16).

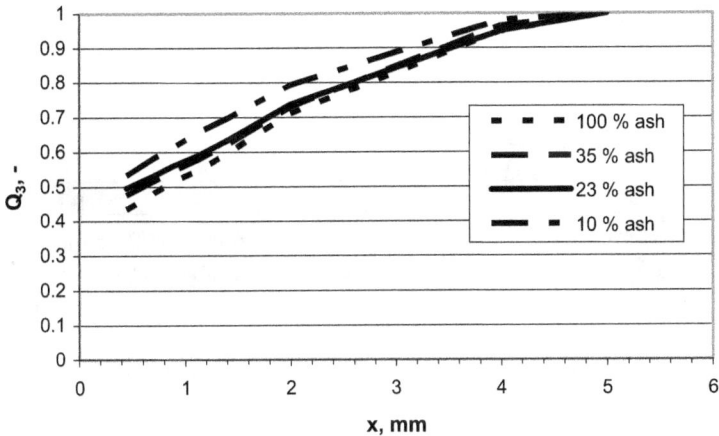

Figure 5-15: Influence of coal ash – sand ratio on PAPSD in fluidization method

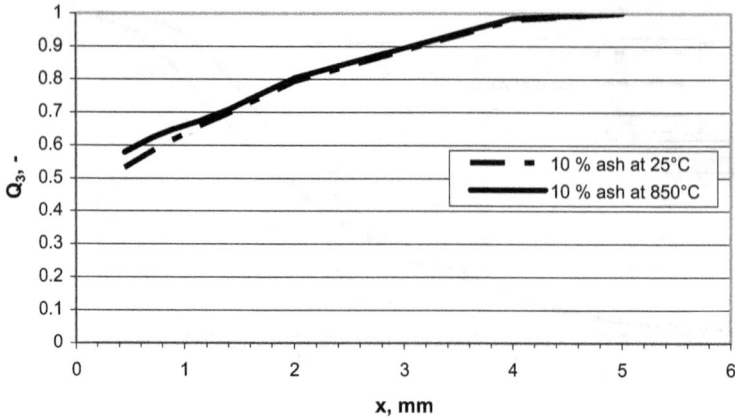

Figure 5-16: Influence of temperature on PAPSD when fluidizing with sand

The results presented above were exclusively conducted with ash particles originating from fuel particles with size in the range of 2 – 5mm. The PSD of the fuel particles is influencing the PAPSD obviously as seen in figure 5-17. The fuel PSD indicated as "full PSD" is the PSD of the fuel particles used in the combustion experiment in the CFBC for model validation and it's PAPSD therefore used as input for the model calculations. The "full PSD" has got a significant amount of fines which will contribute to the fines of the PAPSD. It is remarkable that the PSD of the fine fraction of the PAPSD is the same for both fuel PSDs (cf. figure 5-18) only the amount of fines is different. The reason could be the natural composition and structure of the mineral matrix of the fuel particles.

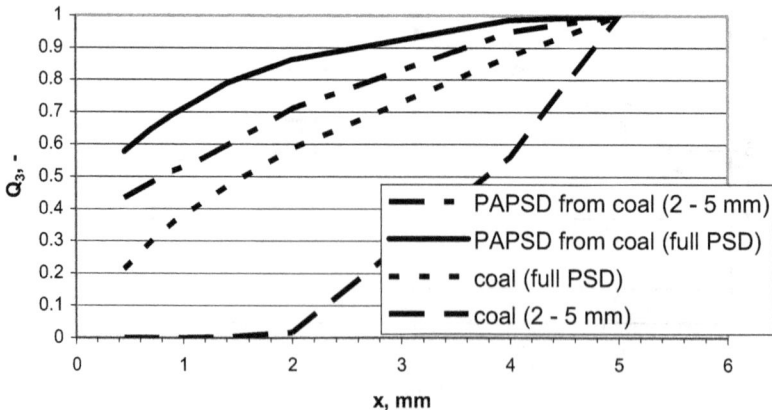

Figure 5-17: Influence of PSD of fuel on PAPSD

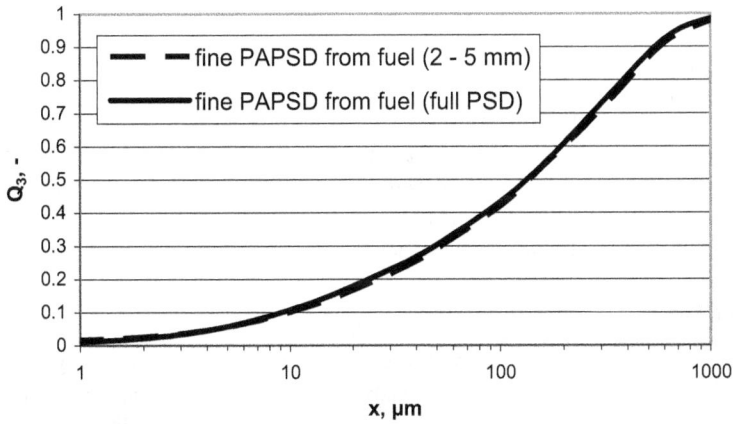

Figure 5-18: Fine fraction of PAPSD from different fuel PSD; fine fractions of PAPSD in Figure 5-17 obtained by sieving at 450 µm.

6. Application of the model

6.1. Batch-wise operating bubbling fluidized bed system

Two different fluidized bed systems with recirculation of solids are calculated with the presented model and compared with experimental data. The first system is a batch-wise operating bubbling fluidized bed which is typical for FCC-reactors. Such a system is ideal to investigate the start-up behavior of the PSD and where it will end after a certain time. In addition it has the advantage that it contains only one kind of solids at the same time.

6.1.1. Model parameters

In the experiments and the calculations two conditions and two different catalysts were applied. One was conducted using spent catalyst which should not undergo the initial attrition like fresh particles. In this case the development of the PSD and the mass loss of the system is only effected by the separation effects of the system and the steady-state attrition of the particles. Contrary to the materials used in the attrition assessments in chapter 5.1 the solids for the tests in a complete system were not sieved before. They were put into the system with the complete PSD as delivered. The PSDs are given in figure 6-1. The spent catalyst is named FCC-A. FCC-B is a fresh catalyst which has never been used in a reactor or other apparatus. It was received directly from the producer.

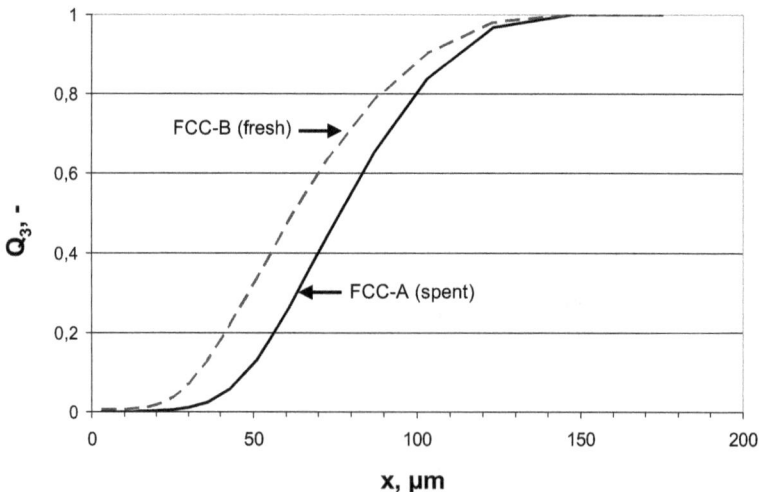

Figure 6-1: Particle size distributions of FCC-catalysts

FCC-B was used for the second test where the impact of the stress history of the particles during start-up is investigated. The solids properties and attrition related parameters which have been determined independently are presented in table 6-1.

Table 6-1: Solids properties for model calculations

	FCC A (spent)	FCC B (fresh)
density, kg/m^3	1550	1560
C_b, s^2/m^4	0.3 10^{-3}	0.53 10^{-3}
C_c, s^2/m^3	0.92 10^{-3}	1.9 10^{-3}
C_j, s^2/m^3	13.1 10^{-6}	12.2 10^{-6}
t_b*, h	-	358
t_j*, h	-	184
$m_{exp,j}$, kg	-	0.2
n_p*, -	-	25
b, -	-	-1.16

6.1.2. Time dependent behavior of spent FCC-catalyst

The test with the spent catalyst FCC-A was started with an initial inventory of the system of 6.5 kg and a fluidizing velocity of 0.3 m/s was applied. Under this condition a velocity of 9 m/s at the cyclone inlet is generated.

The development of the mass flux via the overflow of the cyclone to the filter is presented in figure 6-2. The experimental result shows a high value at the very beginning which drops nearly immediately to a low value. The reason of this effect, which is nearly observed for every run, is not clear. It could be that this is a classification effect which takes place when the gas supply is opened and the flow undergoes a kind of an uncontrolled development in the first moments. Then the mass flux increases over a period of several days to a nearly constant value.

In the calculation two modifications of the cyclone model are tested. The modification of the model by TREFZ & MUSCHELKNAUTZ (1993) which takes a classification at the cyclone entrance into account and is modified to fulfill the mass balances for each size class is shown as the solid line. Using this model the mass flux jumps very fast (within the first 20 minutes) to the constant mass flux which the system reaches after 14 days.

Realizing this discrepancy the model is changed to an older version MUSCHELKNAUTZ (1970) where the classification at the inlet is neglected and the PSD of the material separated at the inlet for strand formation has the same PSD as the material entering the inner vortex which will be consequently the same PSD of material entering the cyclone. Under these circumstances much more fines are collected by the cyclone and the mass flux at the overflow increases much slower. This model is therefore used for all subsequent calculations.

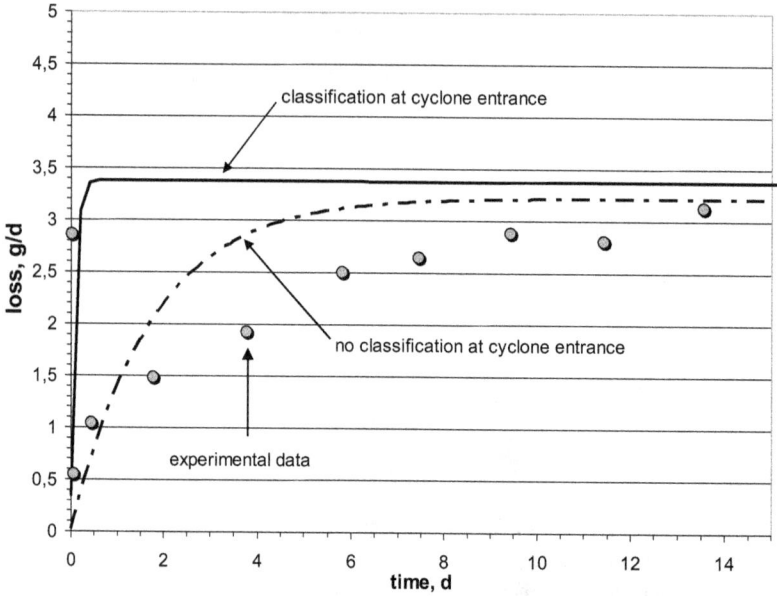

Figure 6-2: Development of PSD for spent material

Figure 6-3 presents the measured and calculated particle size distributions of the bottom bed and of the solids stream in the cyclone entrance. The samples for the PSD of the bottom bed were withdrawn directly through a sample port of the bed and the samples for determining the PSD of the solids entering the cyclone were collected via a suction probe located in the inlet duct to the cyclone as described in chapter 4.3.1, where the sampling procedure is described in detail. The PSD at the cyclone inlet is obtained by averaging 4 different samples taken at 4 different heights in the inlet duct to avoid measurement errors by particle segregation in the inlet flow.

The comparison of the measured with the calculated data shows a fairly good agreement. The development of the PSD of the bottom bed is slightly overestimated by the calculation. This is consistent with the overestimated mass flux at the cyclone overflow.

The integrated modeled higher loss of particles which are in majority fines compared to the measured loss results in a coarser PSD of the remained particles in the bottom bed. This leads in consequence to a coarser PSD of the entrained particles in the cyclone inlet.

The calculations could be adapted to the experimental finding by fitting the model parameters, which is not done here. The calculations are only based on parameters, which were obtained independently in other test or given in literature, respectively.

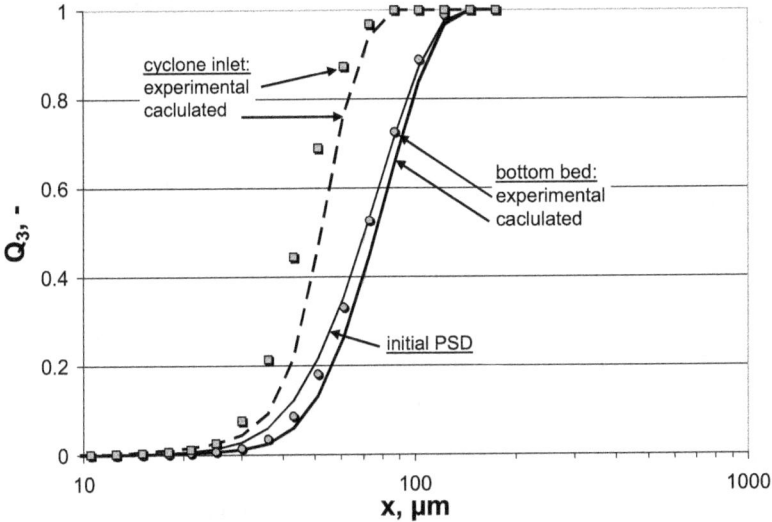

Figure 6-3: Particle size distributions calculated for attrition of spent FCC-catalyst and compared with PSDs measured after 15 days of start.

6.1.3. Time dependent behavior of fresh FCC-catalyst

As a second scenario for a start-up condition a system with fresh particles (FCC-B) is investigated. Here not only separation effects are taking place. The attrition behavior of the particles is changing with time as described in chapter 3.2.2. The particles have not been undergone fluidizing conditions before in the state as delivered. The particles were fluidized at a superficial velocity of 0.45 m/s with cyclone inlet velocity of 13.4 m/s and the same sampling methods were applied as for the test with the spent particles FCC-A.

Figure 6-4 shows the development of the mass loss flux of the system with time. The experimental data show a very high mass loss at the beginning which is decreasing slowly over a period of 8 days to reach a nearly constant rate. To show the effect of the time depending development of the attrition by the change of the stress history the calculation is

85

first conducted with only considering the experienced mechanical stress of the particles by the bubble movement which is represented by the dashed curve.

Then step by step the consideration of the change of stress history by the other sources, namely the jet-induced (dash-dotted line) stress and finally the stress in the cyclone (solid line) shows an increasingly better agreement with the experimental data. Note that the attrition in the sources is calculated at all variations. Only the change in the stress history is differently considered. To be precise: for only considering bubble movement eqns. (3-23) and (3-24) are neglected, where for only considering bubble movement and the jets eqn. (3-23) is neglected and finally for taking the total coupling of all sources into account no equation is neglected and the full model is applied as described in chapter 3.2.2.

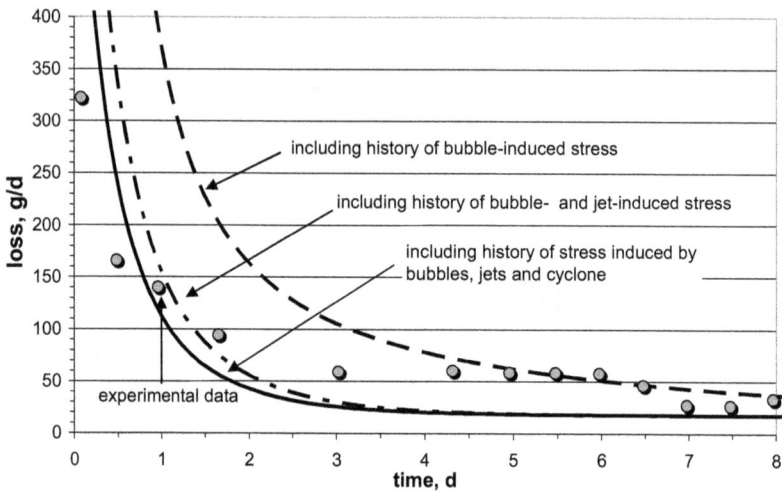

Figure 6-4: Development of mass loss with time for fresh FCC-catalyst B at u = 0.45 m/s.

The results show in a convincing manner that the coupling of the sources has to be considered when dealing with particles which change their behavior with experience of stress. Once again here no fit of parameters and / or introduction of additional parameters is undertaken which would lead to a perfect description of the experimental data. As long as the dependence of such fitting parameters is not known for different solids it is understood that the presented findings should be used to describe in first place the trend of changes in particle size distribution only.

6.2. Circulating fluidized bed combustor (CFBC)

Circulating fluidized bed combustors (CFBC) are a good example of fluidized bed systems involving several different species of solids at the same time. The modeling of such systems has to take the differences in material properties as density and attritability into account. Therefore experiments were conducted at a CFBC pilot plant, which is described in detail in chapter 4.3.2. The results are compared here with calculations.

The model calculations are divided in three sections. First the model is validated with the experimental data of the combustion of dried sewage sludge pellets, coal or a mixture of both, respectively. As a second step the model is applied to investigate the influence of the addition and attrition of limestone for the desulfurization when sewage sludge is combusted. Further applications are then in the third section the application of the model for a load change in the combustion of coal and for the complete change of fuel during operation.

6.2.1. Model Parameters

In all validation experiments quartz sand is used as inert bed material to keep the mass of inventory constant over time. The pressure drop of the riser was kept constant at 7,500 Pa, which corresponds to a riser inventory mass of 5 kg. The experimental runs are simulated with the present model. Input parameters of the model are the dimensions of the CFBC, operating conditions and the determined attrition parameters, which are summarized in table 6-2. As mentioned in chapter 3.1 the feed of fuel is considered only as the feed rate of ash and consequently only the attrition parameters of the ash are needed. The quartz sand is assumed to not undergo attrition.

The quantities C_c, C_j and C_b were directly determined in separate attrition tests for the individual parameters.

Due to a lack of sufficient material amount for attrition assessments with fresh particles some values are approximated and marked with "*" in table 6-2. These latter values were obtained in the following way:

First the values in figure 5-10 were approximated by a set of functions as follows

$$n_p^* = 11.9 \cdot C_c^{-0.12} \qquad \text{for attrition in cyclones} \qquad (6\text{-}1)$$

$$t_b^* = 72.6 \cdot C_b^{-0.24} \qquad \text{for bubble-induced attrition} \qquad (6\text{-}2)$$

$$t_j^* = 135 \cdot C_j^{-0.03} \qquad \text{for jet-induced attrition} \qquad (6\text{-}3)$$

where C_c and C_j have to be inserted in s²/m³ and C_b in s²/m⁴, respectively in order to obtain t_b^* and t_j^* in hours.

It is clear that in a real combustor limestone, calcium oxide and calcium sulfate particles or calcium oxide particles coated with calcium sulfate will coexist. In the present work only limestone is considered in the experimental as well as in the simulation work.

The approximation includes the assumption, that for very small attrition coefficients the smoothening of the particles' will take an infinite time. On the other hand for very high attrition coefficients the limiting value of t_b^*, t_j^* and n_p^* will be zero. The relationships (6-1) - (6-3) are just intended to serve as a first approximation of the characteristic values of t_b^*, t_j^* and n_p^* in cases, where the direct measurement of these values with fresh material is not possible.

Table 6-2: Model parameters of solids for modeling CFBC

	quartz sand	sludge ash	coal ash	limestone
density, kg/m³	2600	1560	2300	2650
C_b, s²/m⁴	0.	61.0 10⁻³	3.94 10⁻³ **	6.0 10⁻³
C_c, s²/m³	0.	34.0 10⁻³	1.22 10⁻³ **	13.5 10⁻³
C_j, s²/m³	0.	9.5 10⁻⁵ *	9.51 10⁻⁶ **	6.82 10⁻⁴
t_b^*, h	-	105	411 *	480
t_j^*, h	-	50 *	190 *	165
$m_{exp.j}$, kg	-	0.25 *	0.25 *	0.25
n_p^*, -	-	17	25 *	22
b, -		-0.7	-0.7 *	-0.7

* approximated values; ** determined with "old" ash from industrial plant

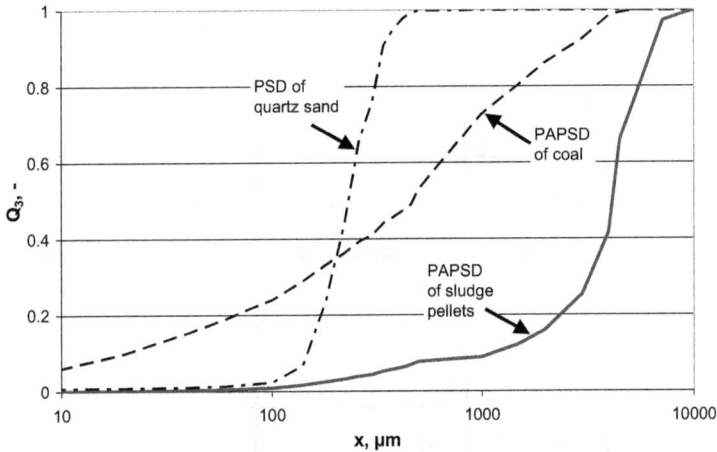

Figure 6-5: PSD of quartz sand and PAPSD of coal and sewage sludge pellets

The particle size distributions of the quartz sand and the fresh ash named PAPSD are given in figure 6-5. The quartz sand has got a very narrow PSD compared to the PAPSD's of the coal and the sewage sludge. The coal ash is finer than the sludge ash. Although the ash of the coal contains a significant higher amount of coarse particles than the sand the amount of fine particles is also much higher. This demonstrates already the big differences of the solids which are likely to have strong influence on the steady-state PSD in the system.

In all experiments the gas velocity at the top of the riser was set to 4.2 m/s and an excess air ratio for the combustion of 1.23, which corresponds to an oxygen concentration of 3.5 vol.-%. This defines the fuel feed rates, which are given in table 6-3.

Table 6-3: Fuel compositions and feed rates

	coal	dried sewage sludge pellets	mixture (50 % coal)
water (raw), kg/kg	0.007	0.15	
ash (wf), kg/kg	0.063	0.366	
ultimate analysis (waf), kg/kg			
carbon	0.81	0.52	
hydrogen	0.05	0.07	
nitrogen	0.02	0.07	
sulfur	0.00	0.02	
oxygen	0.12	0.31	
feed rates, kg/s			
as fuel	$7.27 \; 10^{-4}$	$1.73 \; 10^{-3}$	$1.10 \; 10^{-3}$
ash	$4.55 \; 10^{-5}$	$5.38 \; 10^{-4}$	$2.05 \; 10^{-4}$

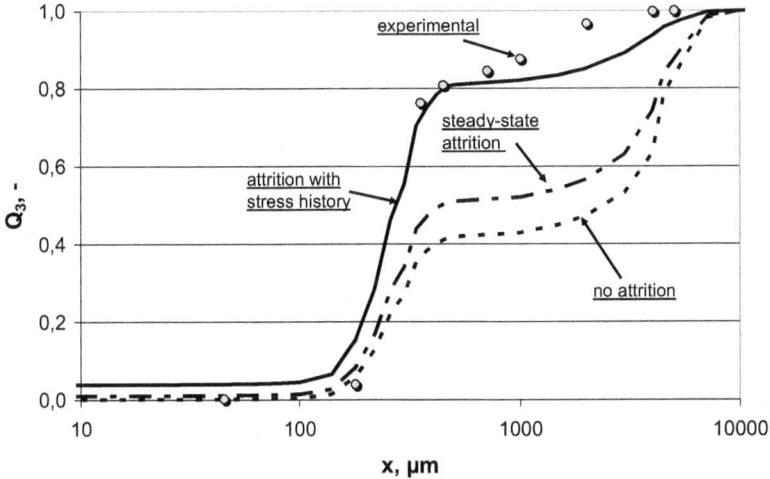

Figure 6-6: PSD in bottom bed for sewage sludge combustion in comparison with different variations of model

6.2.2. Comparison with experimental data

Variation of model calculation

In the model calculation 3 different variations of the modeling are tested. The variations consist of the combination of calculations once without considering the attrition of the solids, then considering the attrition of the solids only by their steady-state attrition behavior and finally by considering their stress history. The results of the 3 variations are presented in figure 6-6. After the results in the previous chapter have shown, that the use of a cyclone model without a classification at the entrance gives better agreement with the experimental findings, the same model has been applied for all the following calculations. The experimental particle size distribution of the bottom bed for combustion of sewage sludge is compared with the calculated ones. When the attrition is not considered in the modeling the fraction of coarse particles is overestimated. These particles are not entrained from the bed and will leave the riser with the bottom drain. The consideration of the steady state attrition yields a particle size distribution, which describes the experiments a little better. But only when the stress history is introduced in the modeling the particle size distribution can be described with a good agreement.

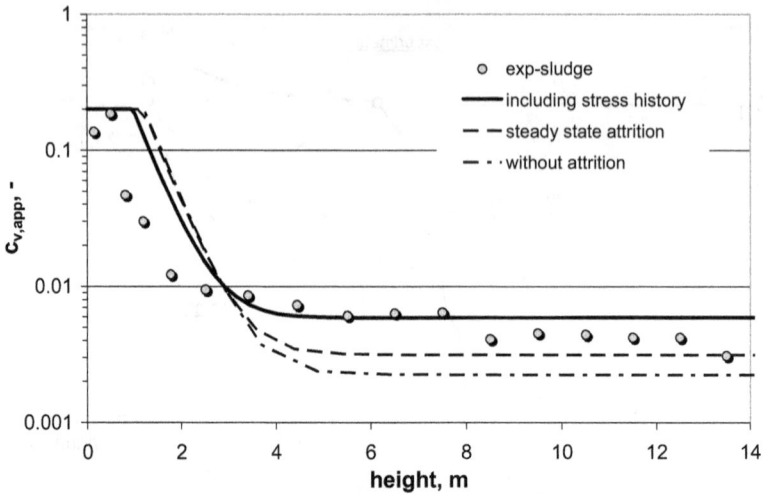

Figure 6-7: Axial solids volume concentration profile while combustion of sewage sludge and model calculations of different variations in model

An important characteristic property of a circulating fluidized bed is its axial solids volume profile, which shows the distribution of material in the riser. Therefore the validation of the model is made by comparison model data with experimental data. The profiles shown in figure 6-7 support the conclusions found in the comparison of the particle size distributions of the bottom bed. When no attrition is considered the solids are accumulated in the bottom bed to a higher extent than with attrition. Attrition leads to a high production rate of fines, which are elutriated from the bottom bed and transported to the top of the riser. Again, considering the attrition including the stress history yields a good agreement between simulation and experimental values.

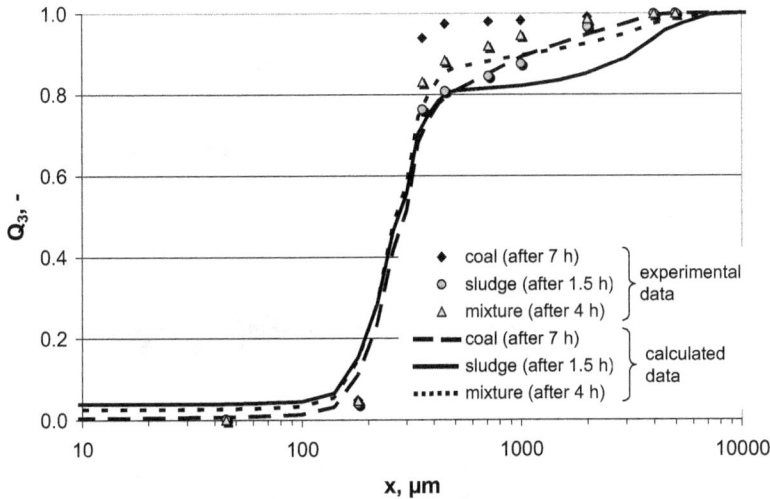

Figure 6-8: The particle size distributions in the bottom bed for different fuels

Variation of fuel

Different fuels with different ash composition or ash particle sizes and attrition behavior are generating different particle size distributions of the bottom ash, which is presented in figure 6-8. The use of coal as the fuel leads to a much finer bottom ash due to its finer PAPSD (cf. figure 6-5). The 8 times lower feed rate of coal ash compared with the feed rate of sludge ash causes a higher consumption of sand to keep the inventory mass constant. The feed rate of fuel is decided by the gas velocity at the top of the riser and the desired excess air ratio. Therefore the PSD of the bottom bed is more dominated by the quartz sand when coal with low ash content is applied. When a mixture of coal and sewage sludge of about 50% each as fuel, which results in 86% of sludge ash, the PSD of the bottom bed lies in between of the PSDs, when pure fuels are used. Figure 6-8 shows the qualitative agreement of the model calculation for all fuels with the experimental data. However, the model assumes that the attrition behavior of the particles is independent of the presence of particles from different materials. This is certainly a simplifying assumption. It is very probable, that the more friable ash particles will be easier attrited in combination with the attrition-resitant quartz sand. This will lead to general underestimation of the attrition rates.

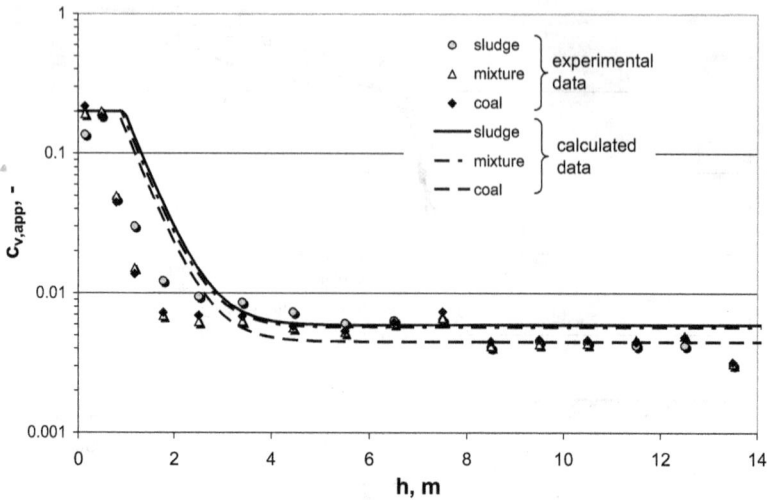

Figure 6-9: The axial solids volume concentration profiles in the riser of a circulating fluidized bed combustor for different fuels

The comparison of the calculated axial solids volume concentration profiles for all fuels with the experimental data confirms the good agreement of the model (cf. figure 6-9). The differences in the solids volume concentrations are hard to detect in the experiment. The differences are very small and are smaller than the natural fluctuations of the system. In the calculations the differences are easier to see. It has to be remarked, that in contradiction to a real system, in the calculation an ideal operation is modeled without any scattering. But circulating fluidized beds are characterized by their strong fluctuations, which makes it difficult to keep the conditions constant in a narrow range, and which would make it thus impossible to detect such small differences in the solids concentration. Taking a closer look at figure 6-9 shows that in the section below 5 m above the distributor the concentrations of the measured volume concentrations for the sludge combustion is higher than for the combustion of coal, which is due to the coarser bottom bed PSD in the sludge case. Since the riser inventory is the same in both cases the solids volume concentration during coal combustion is then higher than for sludge combustion in the upper section of the riser.

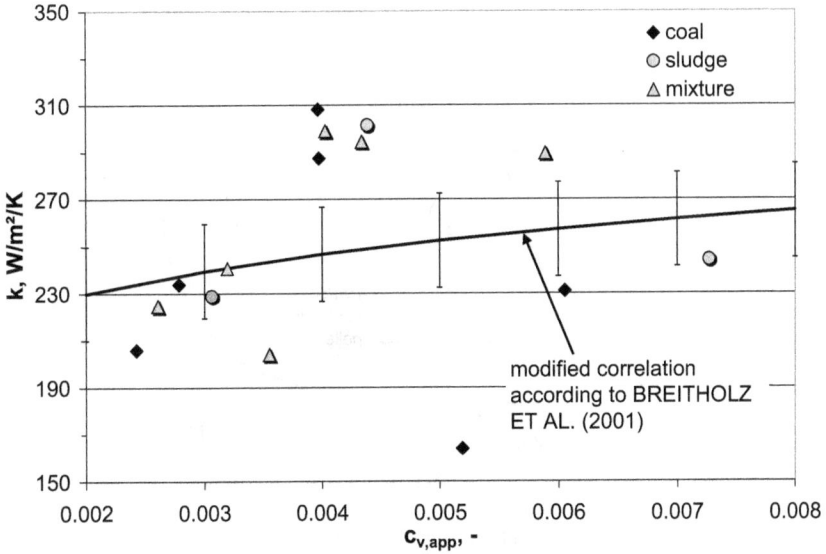

Figure 6-10: Heat transfer coefficient as function of solids volume concentration

To detect changes in the bed to wall heat transfer with changes in solids volume concentration a heat transfer probe was used at different locations in the riser. Figure 6-10 shows the dependence of the measured heat transfer coefficients on the solids volume concentration. In spite of a considerable scatter it can be stated that the heat transfer coefficient increases slightly with increasing concentration of solids. An influence of the kind of fuel cannot be concluded. Only the amount of solids in the vicinity of the heat transfer surface is decisive for the transferred heat, but not the composition of the particles.

Since a horizontal pipe was used as heat transfer surface in the present work a correlation suggested by BREITHOLZ et al. (2001) for the heat transfer to vertical membrane walls, was modified to describe the experimental in this work. This new correlation is

$$k = 192.2 \cdot \rho_{sus}^{0.1} \; \frac{W}{m^2 \cdot K} \tag{6-4}$$

when the suspension density ρ_{sus} is inserted in kg/m^3.

The suspension density ρ_{sus} can be calculated from the solids volume concentration c_v by

$$\rho_{sus} = c_v \cdot \rho_s + (1 - c_v) \cdot \rho_f \tag{6-5}$$

95

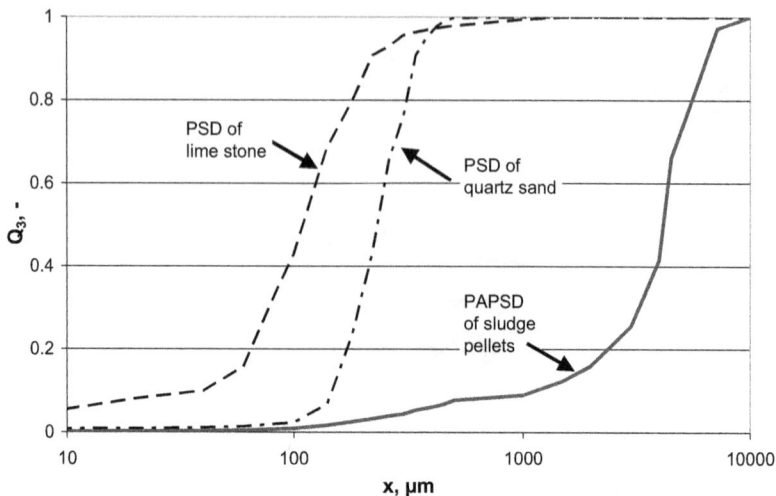

Figure 6-11: PSD of limestone and quartz sand and PAPSD of sewage sludge pellets

6.2.3. Influence of sorbent on PSD

Emission control is a very important issue in industrial processes and therefore also in combustion. One example in combustion processes is the emission of sulfur dioxide SO_2 when fuels with high sulfur contents are combusted. In order to deal with this problem and to minimize the emission of SO_2 the addition of limestone is a successful method. The limestone reacts first to CaO which subsequently reacts with the sulfur and generates gypsum which can be collected together with the fly ash or bottom ash. For the reaction of one mole of sulfur one mole of limestone is needed on a stoichiometric basis. But the experience from industrial application shows that an excess of limestone to sulfur content has to be used in order to achieve the desired emission limits in the exhaust flue gas. Therefore it is of high interest to know how the sorbent limestone is distributed in the system and what residence times it has or which impact the addition of sorbents has on the performance of the process, respectively. The addition of the sorbent will have an influence on the PSD in the bed, too.

The addition of limestone is applied to investigate its possible impact on the bed particle size distribution. Here two cases are calculated on the basis of the operating conditions of the above presented experiments with the combustion of the dried sewage sludge pellets which have very high sulfur content. In the first case a calcium-to-sulfur ratio on a molar basis of 1 is used which will mean that in the calculation a mixture of 10%limestone with 90% sludge-ash at a feed rate of $5.98 \cdot 10^{-4}$ kg/s is applied. In the second case a ratio of 10 is calculated which is a high excess of limestone but not unusual in industry. In this latter case $11.4 \cdot 10^{-4}$ kg/s are fed containing 53% limestone and 47% sludge-ash. The particle size distributions of the solids are compared in figure 6-11.

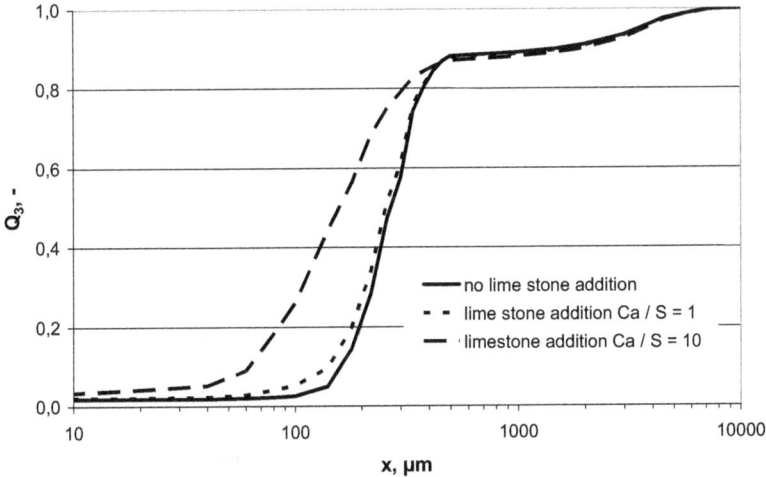

Figure 6-12: The influence of limestone addition on particle size distribution of bottom bed.

The addition of the limestone leads to an increase of fines in the bottom bed of the combustion chamber as is shown in figure 6-12. Looking at the PSD of the limestone, which is much finer than the one of the sludge PAPSD (cf. Figure 6-11) this is not a surprising result. However, for the Ca/S-ratio of 10 a very significant impact on the PSD is observed.

The consequences become clearer when the distribution of solids within the combustion chamber is seen in figure 6-13. Keeping the total mass of inventory in the combustion chamber constant the addition of limestone leads to a significant decrease of the bottom bed and as a consequence to an increase of solids volume concentration of the upper dilute region. This will lead to higher heat transfer rate in the upper dilute region. Eqn. (6-4) gives for the the pure sludge combustion a heat transfer coefficient k of 340 W/m²/K, which increase from the stoechiometric addition of limestone to 342 W/m²/K to 352 W/m²/K for the 10 times higher addition of limestone than stoichiometry. In addition, the solids concentration in the flue gas will increase and this has to be considered in the design of the gas cleaning system.

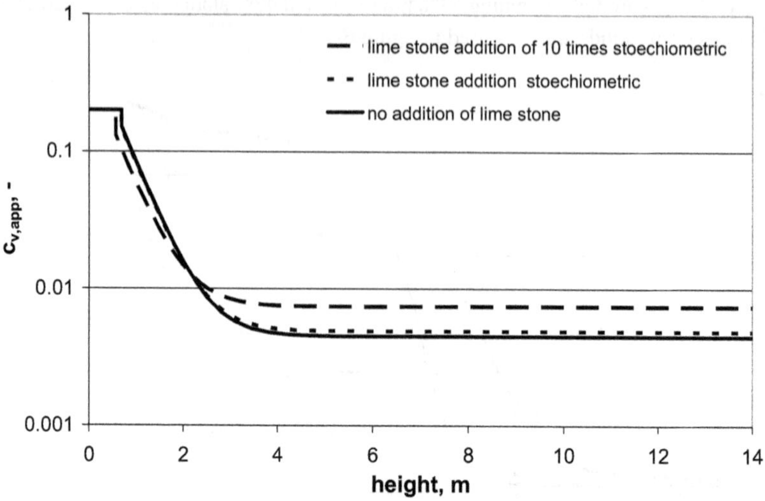
lime stone addition of 10 times stoechiometric
lime stone addition stoechiometric
no addition of lime stone
$c_{v,app}$, -
1
0.1
0.01
0.001
0
2
4
6
8
10
12
14
height, m

Figure 6-13: The influence of limestone addition on axial solids volume concentration profile.

6.2.4. Application of the model to a load change

After validation the model was used to investigate the temporal development of the particle size distribution and the vertical solids concentration profile as a consequence of a change in operating conditions. As an example in the following simulation results for a stepwise load change are given. In this case the load has been increased after 10 hours at a load of 60 % to full load. As 60% load the conditions of the validation experiments for coal combustion was taken with the gas velocity of 4.2 m/s and an ash feed of $4.55 \ 10^{-5}$ kg/s. The full load case is then defined by an ash feed rate of $7.43 \ 10^{-5}$ kg/s and a gas velocity of 7 m/s to keep the excess air ratio constant with 1.23.

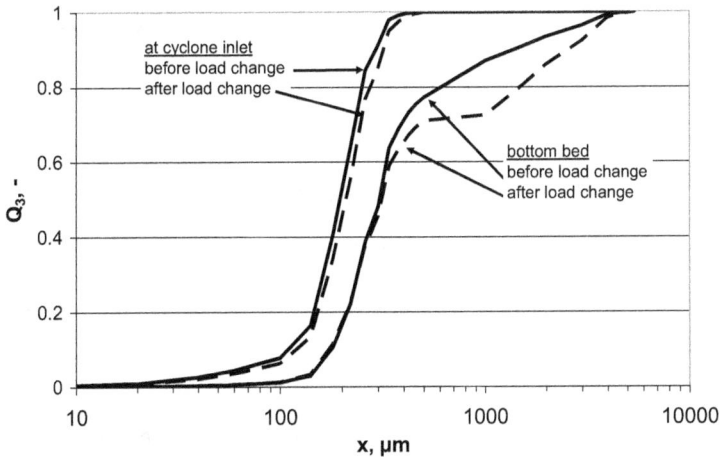

Figure 6-14: Particle size distributions in the bottom bed and cyclone inlet before (10h after start) and after the load change (20h after start). The load change in form 60 to 100%.

Figure 6-14 illustrates the influence of the load change on the particle size distrubtions at the cyclone inlet and in the bottom bed. At both locations the PSD's are getting coarser. Figure 6-15 shows the development of the surface mean diameters of the bottom bed and the circulating material at the cyclone inlet. The first ten hours in this plot depict the development of the average particles size with time beginning with the start-up conditions for the 60 % load case. The average diameters of both materials increase suddenly at the time of the load change (10h after start). This increase is a result of the suddenly increased gas velocity, which allows larger particles to be carried out. Especially entrainment of particles in the size range between 100 and 250 μm will rise spontaneously. This higher amount of particles larger than 100 μm in the circulating material causes the increase of the average diameter of the circulating material in the cyclone inlet. On the other hand by this enlarged entrainment the hold-up of particles with diameters less than 250 μm in the bottom bed is decreased, which again causes an increase of the average diameter of the bottom bed. Figure 6-16 illustrates the influence on the axial profile of the solids volume concentration.

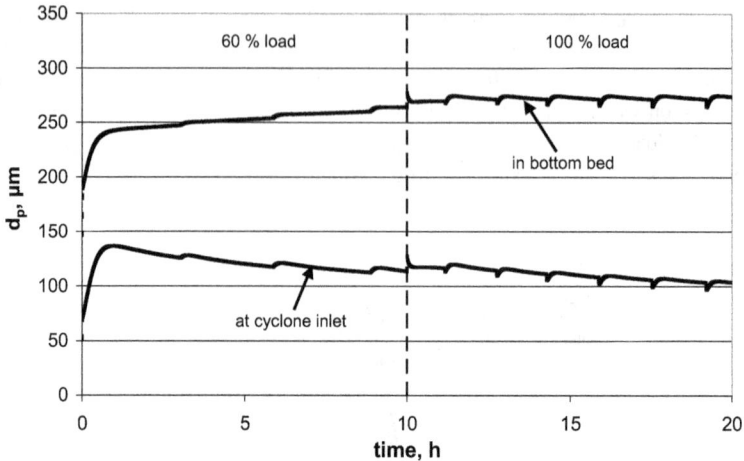

Figure 6-15: Simulation of a load change: development of the mean surface diameter of the circulating material and the bottom bed material

It is clearly noticeable that the increase of the particle size in the bottom bed with time is stopped for the 100 % load while the particle size at the cyclone inlet is still decreasing with time. This is mainly due to an increase of the fines produced by attrition.

From Figure 6-16 it can be derived that the heat transfer in the upper part of the riser will increase as long as due to the increased gas velocity the solids volume concentration increases by a factor of nearly 2. According to eqn. (6-4) the heat transfer coefficient will increase from 242 W/m²/K to 256 W/m²/K by 6%.

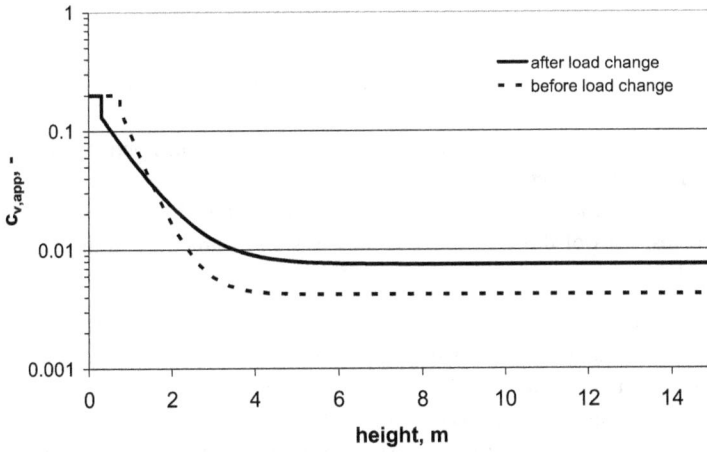

Figure 6-16: Axial solids volume concentration profiles before (10h after start) and after load change (20h after start) from 60 to 100%.

7. Summary and Conclusions

In the present work an approach has been developed to predict the changes in particle size distribution of solids in fluidized bed systems with recirculation of solids. Several effects in fluidized bed systems are influencing the local particle size distribution and have been identified. They are grouped in separation effects, transport effects and degradation effects of the individual particle size. Since a lot of fluidized bed applications are dealing with mixtures of particles of different materials this was also introduced in the scope of the present work.

For separation and transport effects a lot of models are given in the literature, but usually only applicable to the description of systems of homogeneous materials. For the modeling of systems containing solids of different materials they had to be adapted. The models have to be able to distinguish between particles of different densities. The effects, which had to be modeled, are the elutriation of particles from a fluidized bed and the separation in gas cyclones. For the elutriation of particles the terminal velocity of the particles is decisive, which is a main parameter in the published models. A modification was therefore not necessary only the mass ratio of the different materials in the mixture is introduced as a correction factor.

The widely used Muschelknautz model for separation in cyclones assumes a constant density for all the particles despite the fact that they are separated according to their terminal velocity in the flow field of the vortex. In the present work the characteristic particle sizes are replaced by characteristic or critical terminal velocities, which have been calculated for the centrifugal forces and drag forces in the vortex flow field. In the Muschelknautz model two separation mechanisms are distinguished, which are working subsequently. When the particles are entering the cyclone and their concentration in the gas flow is above a critical value, all particles above this value are separated spontaneously at the inlet and form a strand, which flows down the cyclone inner wall. The remaining particles in the gas flow are entering the inner vortex and are separated depending on their size and density in the centrifugal field. Up to now it is not quite clear if there is a classification taking place during the first separation at the inlet or not. Therefore both variations, one with classification and one without, were tested in the modeling. For the systems investigated in the present work it turns out, that neglecting a classification effect at the inlet leads to much better agreement with experimental data.

While the separation effects are only depending on the momentary conditions and do not depend on what has happened in the past, this is not true for the attrition of the particles. Attrition has a strong effect on the particles sizes, especially when fresh particles are applied, which have not undergone mechanical stress before. In previous investigations it has been observed that fresh particles have a much higher production rate of fines than particles, which have been exposed to attrition stress for a long time. The attrition rate of fresh particles can be 2 to 3 times in order of magnitude higher than the attrition rate measured after a long time of exposure. This effect is referred to the condition of the particles surface. Fresh particles have a very rough surface with a lot of edges and small cracks. After a certain period of time exposed to abrasion condition these edges are rubbed off and the particle surface gets very smooth. Then the attrition rate is defined by a natural roughness of the particles, which is depending on the inner structure of the individual particle.

From this finding it is concluded, that the stress experienced by the particles has to be monitored and taken into account in the modeling. This experienced stress is called here the "stress history" of the particles. In fluidized bed systems with recirculation of solids the stress history is affected by three sources of attrition, namely attrition induced by jets at the gas distributor in the fluidized bed, attrition induced by bubble movement in the bed itself and attrition in cyclones. Whenever a particle is exposed to one of these sources its stress history is changed and this will influence the attrition behavior when it is subsequently subjected to another attrition mechanism. This illustrates a strong coupling of these sources. Even when two particles have been for the same time in the system they may have undergone a different stress history. For example smaller particles have been recirculated and stressed in the cyclone more often than coarser ones.

The approach chosen in the present work is to define a dimensionless stress history parameter, which defines the status of a particle at a certain moment. Stress history parameter $\vartheta < 1$ means, that a particle is not yet in a steady state condition regarding the attrition. For $\vartheta > 1$ the particle is in a state, where it will have the attrition behavior of a fully smoothed particle. The attrition of particles with $\vartheta > 1$ is therefore described by the models obtained in previous investigations, which were focused on the attrition in the steady state. The attrition rates of particles with $\vartheta < 1$ are calculated by applying a factor $f(\vartheta)$ which is depending on ϑ in which the information of the stress history is saved. The change of the stress history during stressing procedure is then implemented in a model, depending on the source of stress. In the fluidized bed this change depends on the time of exposition and in the cyclone it is a stepwise change for every pass through the cyclone. Where the attrition behavior and coefficients are different for solids from different materials, this applies for the dependence on the stress history represented by $f(\vartheta)$ too. For the whole description of the attrition induced by one source a set of two coefficients is needed, namely the attrition coefficient and a reference for defining ϑ to be equal 1, i.e. time or number of passes, respectively. For the unification of the description for all sources one parameter is needed to describe the shape of $f(\vartheta)$. This parameter can be derived from experiments with isolated sources of attrition.

In order to describe a complete, arbitrary system of combinations of fluidized beds and cyclones the model system is divided into modules, i.e. one for each apparatus in the system. Each module is calculating based on the previously mentioned models the changes in particle size distribution, stress history and the mass fluxes and particle size distribution at the outlets of the module, which is then the input to the subsequent module. This modeling strategy is then applied to two systems consisting of a fluidized bed, a cyclone and a recirculation of the underflow of the cyclone back into the fluidized bed, at which experiments have been conducted and data were available. The first system is a bubbling fluidized bed with FCC-catalyst particles as inventory, which is operating batch-wise. Here the start-up behavior of a system, when initialized with fresh material is investigated. A circulating fluidized bed combustor is chosen as second system, which is characterized by much higher fluidizing velocities, continuously fed with fresh material and containing solids of more than one kind.

At first for all the materials the model related parameters were obtained from experiments at isolated and independent experimental set-ups or taken from literature. With these parameters the system was simulated and the results from calculations compared with experimental findings without any further parameter fitting. The calculations meet with a reasonable good agreement and accuracy the experimental data. Changes in particle size distributions and mass fluxes with time are described qualitatively very good. Inaccuracies are obtained on one hand by inaccuracies in the measurements and on the other hand by assumptions made in the model. The simplification was made that with regard to attrition the particles of different material are not influencing each other. This had to be made as a first simplification being aware, that this is the reality. It can easily be imagined that the attrition of a friable particle is different when it is mixed with particles of a more attrition resistant material.

However, the model is then applied to different scenarios, which are occurring during operation of a circulating fluidized bed combustor, e.g. addition of sorbent for emission control and load change. From the application the trends of changes in particle size distribution, mass fluxes and the depending operating characteristics, e.g. heat transfer can be derived. The calculations illustrate that the simulation tool derived in the present work may be useful for practical applications in industry.

References

Bellgardt D., M. Schößler and J. Werther, "Lateral non-uniformities of solids and gas concentrations in fluidized bed reactors", Powder Technol., 53, pp. 205-216, 1987.

Boëlle A., M. Qian, P. Jaud, R. Chirone, P. Salatino, F. Winter, X. Liu, D. Olsson, L. Amand and B. Leckner, "Coal comminution characterization for industrial scale Circulating fluidized bed", Final Joint Report Electricité de France, 2002.

Brauer H., "Grundlagen der Einphasen- und Mehrphasenströmungen", Aarau, Frankfurt a.M.: Sauerländer, 1971.

Breitholtz C., B. Leckner and A.P. Baskakov, "Wall average heat transfer in CFB boilers", Powder Technol., 120, pp. 41-48, 2001.

British Materials Handling Board, "Particle Attrition", Trans. Tech. Publications Series on Bulk Materials Handling, 5, 1987.

Cammarota A., R. Chirone, A. Marzocchella and P. Salatino, "The relevance of attrition phenomena to the establishment of solids inventory and particle size distribution in circulating fluidized bed coal combustors", in: J.R. Grace, J.J.-X. Zhu, H. de Lasa (Eds.), Circulating Fluidized Bed Technology VII., Canadian Society for Chemical Engineering, Ottawa, Canada, p. 661 – 668, 2002.

Cammarota A., R. Chirone, A. Marzocchella and P. Salatino, "Assessment of ash inventory and size distribution in fluidized bed coal combustors", in: Proc. of Int. Conf. on FBC, Reno (NV), NSME (NY), paper 78, 2001.

Cheng L., B. Chen, N. Liu, Z. Luo and K. Cen, "Effect of characteristic of sorbents on their sulfur capture capability at a fluidized bed condition", Fuel, 83, pp. 925 – 932, 2004.

Clift R. and J.R. Grace, "Bubble interaction in fluidized beds", Chem. Eng. Progr. Symp Ser., 66 (105), pp. 14 – 27, 1970.

Clift R. and J.R. Grace, "Bubble coalescence in fluidized beds: comparison of two theories", AIChE J., 17 (1), pp. 252 - 254, 1971.

Colakyan M. and O. Levenspiel, "Elutriation from fluidized beds", Powder Technol., 38, pp. 223 - 232, 1984.

Davidson J.F. and D. Harrison, "Fluidised particles", Cambridge: Cambridge University Press, 1963.

De Vries R.J., W.P.M. van Swaaij, C. Mantovani and A. Heijkoop, "Design criteria and performance of the commercial reactor for the shell chorine process", In: Proc. of the 5th Europ. Symp. on Chem. React. Engg., Amsterdam, The Netherlands, B9-59, 1972.

Dessalces G., F. Kolenda and J.P. Reymond, "Attrition evaluation for catalysts used in fluidized or circulating fluidized bed reactors", AIChE: Preprints of the First International Particle Technology Forum, Part II, Denver, Colorado, pp. 190 - 196, 1994.

Di Benedetto A. and P. Salatino, "Modeling attrition of limestone during calcinations and sulfation in a fluidized bed reactor", Powder Technol., 95, pp. 119 – 128, 1998.

Do HT., JR. Grace and R. Clift, "Particle ejection and entrainment from fluidised beds", Powder Technol., 6, pp. 195 - 200, 1972.

Forsythe W.L. and W.R. Hertwig, "Attrition characteristics of fluid cracking catalysts", Ind. Eng. Chem., 41, pp. 1200 - 1206, 1949.

Geldart D., "Types of gas fluidization", Powder Technol., 7, pp. 285-292, 1973.

Gwyn J.E., "On the particle size distribution function and the attrition of cracking catalysts", AIChE-J., 15, pp. 35 - 38, 1969.

Heidenhof N. and W. Althoff, "Importkohleeinsatz in der zirkulierenden atmosphärischen Wirbelschichtfeuerung des Heizkraftwerkes I der Stadtwerke Duisburg AG", VGB PowerTech, 12, pp. 87 – 89, 2003.

Herbertz H.A., H. Vollmer, J. Albrecht and G. Schaub, "Die zirkulierende Wirbelschicht als Feuerungssystem für Brennstoffe mit hohen und schwankenden Aschegehalten", VGB Kraftwerkstechnik, 69 (10), pp. 1003 – 1008, 1989.

Hilligardt K. and J. Werther, "Influence of temperature and properties of solids on the size and growth of bubbles in gas fluidized beds", Chem. Eng. Technol., 10, pp. 272 – 280, 1987.

Hugi E. and L. Reh, "Design of cyclones with high solids entrance loads", Chem. Eng. Technol., 21 (9), pp. 716 - 719, 1998.

Kaskas A., "Berechnung der stationären und instationären Bewegung von Kugeln in ruhenden und strömenden Medien", Diplomarbeit am Lehrstuhl für Thermodynamik und Verfahrenstechnik der Technischen Universität Berlin, 1964.

Krambrock W., "Die Berechnung des Zyklonabscheiders und praktische Gesichtspunkte zur Auslegung", Aufbereitungstechnik, 12 (3), pp. 391 - 401, 1971.

Kunii D. and O. Levenspiel, Fluidization Engineering. Butterworth-Heinemann, Stoneham, USA, 1991.

Lalak I., J. Seeber, F. Kluger and St. Krupka, "Operational experience with high efficiency cyclones: comparision between boiler A and B in the Zeran Power Plant, Warsaw/Poland", VGB Power Tech, 9, pp. 90-94, 2003.

Leith D. and W. Licht, "The collection efficiency of cyclone type particle collectors - a new theoretical approach", AIChE Symposium Series, 68 (126), pp. 196 - 206, 1972.

Levy EK., H.S. Caram, J.C. Dille and S. Edelstein, "Mechanisms for solids-ejection from gas-fluidized beds", AIChE Journal, 29, pp. 383 - 388, 1983.

Lewis WK., E.R. Gilliland and P.M. Lang, „Entrainment from fluidized beds", Chem. Eng. Prog. Symp. Ser., 58, pp. 65 - 72, 1962.

Molerus O., "Heat transfer in gas fluidized beds", Powder Technology, 70, pp. 1 – 14, 1992.

Mothes H. and F. Löffler, "Zur Berechnung der Partikelabscheidung in Zyklonen", Chem. Eng. Process., 18, pp. 323 - 331, 1984.

Muschelknautz U. and E. Muschelknautz, "Improvements of cyclones in CFB power plants and quantitative estimations of their effects on the boilers solids inventory", in: Circulating Fluidized Bed Technology VI (Ed. J. Werther), Dechema, Würzburg, pp. 761 – 767, 1999.

Muschelknautz U. and E. Muschelknautz, "Special design of inserts and short entrance ducts to recirculating cyclones", Proceedings of the Fifth International Conference on Circulating Fluidized Beds, Beijing, P. R. China, 28.5. - 31.5., pp.597-602, 1996.

Muschelknautz E., "Auslegung von Zyklonabscheidern in der technischen Praxis", Staub-Reinhalt. Luft, 30, pp. 187 - 195, 1970.

Obermair St. and G. Staudinger, "The dust outlet of a gas cyclone and its effects on separation efficiency", Chem. Eng. Technol., 24, pp. 1259 - 1263, 2001.

Pell M. and S.P. Jordan, "Effects of fines and velocity on fluidized bed reactor performance", AIChE Symp. Ser., 84 (262), pp. 68 – 73, 1988.

Pis J.J., A.B. Fuertes, V. Artos, A. Suarez, and F. Rubiera, "Attrition of coal and ash particles in a fluidized bed", Powder Technol., 66, pp. 41 – 46, 1991.

Pratchett T., "Echt Zauberhaft", Goldmann, München, 1997.

Rangelova J., L. Mörl, S. Heinrich and J. Dalichau, "Decay behavior of particles in a fluidized bed – application of a mass-related attrition coefficient", Chem. Eng. Technol., 25 (6), pp. 639 - 646, 2002.

Rangelova J., L. Mörl, S. Heinrich and R. Peters, "Zerfallsverhalten von Partikeln in Wirbelschichten – Anwendung eines konstanten oberflächenbezogenen Abriebskoeffizienten", Chemie Ingenieur Technik 76 (8), pp. 1078 - 1086, 2004.

Reppenhagen J. and J. Werther, "Catalyst attrition in cyclones", Powder Technol. 113, pp. 55 – 69, 2000.

Reppenhagen J. and J. Werther, "The role of catalyst attrition in the adjustment of the steady-state particle size distribution in fluidized bed systems", in: M. Kwauk, J. Li, W.-C. Wang (Eds.), Proc. 10th Eng. Found. Conf. Fluidization. United Engineering Foundation, New York, pp. 69-76, 2001.

Reppenhagen J., A. Schetzschen and J. Werther, "Find the optimum cyclone size with respect to the fines in pneumatic conveying systems", Powder Technol., 112, pp. 251-255, 2000.

Reppenhagen J., "Catalyst attrition in fluidized bed systems", Dissertation TUHH, Shaker, Aachen, 2000.

Scarlett B., "Particle populations - to balance or not to balance, that is the question!", Powder Technol., 125, pp. 1-4, 2002.

Sit S.P. and J.R. Grace, "Effect of bubble interaction an interphase mass transfer in fluidized beds", Chem. Eng. Sci., 36, pp. 327 - 335, 1981.

Sundaresan R. and A.K. Kolar, "Core heat transfer studies in a circulating fluidized bed", Powder Technol., 124, pp. 138-151, 2002

Tasirin SM and D. Geldart, "The entrainment of fines and superfines from fluidized beds", Powder Handling & Processing, 10, pp. 263 - 268, 1998.

Tasirin SM, and Geldart D. Entrainment of FCC from fluidized beds - a new correlation for the elutriation rate constants", Powder Technol., 95, pp. 240 - 247, 1998b.

Trefz M. and E. Muschelknautz, "Extended cyclone theory for gas flows with high solids concentrations", Chem. Eng. Technol. 16 (3), pp. 153-160, 1993.

Weeks S.A., and P. Dumbill, "Method speeds FCC catalyst attrition resistance determinations", Oil & Gas Journal, 88, pp. 38 – 40, 1990.

Werdermann C.C., „Feststoffbewegung und Wärmeübergang in zirkulierenden Wirbelschichten von Kohlekraftwerken", Dissertation TUHH, Shaker, Aachen, 1993.

Werther J., "Bubble growth in large diameter fluidized beds", In Fluidization Technology; Hemisphere Publ.: Washington, pp. 215-235, 1976.

Werther J. and E.-U. Hartge, "Elutriation and entrainment", In: Handbook of Fluidization and Fluid-Particle Systems (W.-C. Yang, ed.), Marcel Dekker, New York, pp. 113-128, 2003.

Werther J. and J. Reppenhagen, "Catalyst attrition in fluidized-bed systems", AICHE J., 45 (9), pp. 2001 – 2010, 1999.

Werther J. and J. Wein, "Expansion behavior of gas fluidized beds in the turbulent regime", AIChE. Symp. Ser., 90 (301), pp. 31–44, 1994.

Werther J. and W. Xi, "Jet attrition of catalyst particles in gas fluidized beds", Powder Technol., 76, pp. 39 – 46, 1993.

Werther J. and E.-U. Hartge, "Modeling of industrial fluidized bed reactors", Ind. Eng. Chem. Res., 43, pp. 5593 - 5604, 2004.

Whitcombe J.M., I.E. Agranovski and R.D. Braddock, "Attrition due to mixing of hot and cold FCC catalyst particles", Powder Technol., 137, pp. 120 – 130, 2003.

Xi W., "Katalysatorabrieb in Wirbelschichtreaktoren", Dissertation TUHH, Shaker, Aachen, 1993.

Zenz PA and N.A. Weil, „A theoretical-empirical approach to the mechanism of particle entrainment from fluidized beds" AIChE J., 4, pp. 472 - 479, 1958.